Schott Guide to Glass

Schott Guide to Glass

Heinz G. Pfaender
Former Professor of Industrial Design
at the Fachhochschule Darmstadt, Darmstadt, Germany

CHAPMAN & HALL
London · Glasgow · Weinheim · New York · Tokyo · Melbourne · Madras

Published by Chapman & Hall, 2–6 Boundary Row, London SE1 8HN, UK

Chapman & Hall, 2–6 Boundary Row, London SE1 8HN, UK

Blackie Academic & Professional, Wester Cleddens Road, Bishopbriggs, Glasgow G64 2NZ, UK

Chapman & Hall GmbH, Pappelallee 3, 69469 Weinheim, Germany

Chapman & Hall USA, 115 Fifth Avenue, New York, NY 10003, USA

Chapman & Hall Japan, ITP-Japan, Kyowa Building, 3F, 2-2-1 Hirakawacho, Chiyoda-ku, Tokyo 102, Japan

Chapman & Hall Australia, 102 Dodds Street, South Melbourne, Victoria 3205, Australia

Chapman & Hall India, R. Seshadri, 32 Second Main Road, CIT East, Madras 600 035, India

English language edition 1996

© 1996 Chapman & Hall

Original German language edition Schott Glaslexikon – © 1989, mvg – Moderne Verlagsgesellschaft mbH

Typeset in 10/12 pt Palatino by Keyset Composition, Colchester, Essex
Printed in Great Britain at the University Press, Cambridge

ISBN 0 412 71960 6 (HB)   0 412 62060 X (PB)

Apart from any fair dealing for the purposes of research or private study, or criticism or review, as permitted under the UK Copyright Designs and Patents Act, 1988, this publication may not be reproduced, stored, or transmitted, in any form or by any means, without the prior permission in writing of the publishers, or in the case of reprographic reproduction only in accordance with the terms of the licences issued by the Copyright Licensing Agency in the UK, or in accordance with the terms of licences issued by the appropriate Reproduction Rights Organization outside the UK. Enquiries concerning reproduction outside the terms stated here should be sent to the publishers at the London address printed on this page.

The publisher makes no representation, express or implied, with regard to the accuracy of the information contained in this book and cannot accept any legal responsibility or liability for any errors or omissions that may be made.

A catalogue record for this book is available from the British Library

Library of Congress Catalog Card Number: 95-69611

∞ Printed on acid-free text paper, manufactured in accordance with ANSI/NISO Z39.48-1992 (Permanence of Paper).

Contents

Foreword xi
Introduction xiii

1 The history of glass 1
  1.1 Glass in Egypt 1
  1.2 A revolution in technology: the glassblowing pipe 3
  1.3 Glass in the period of the Roman Empire 4
  1.4 From luxury product to everyday item 5
  1.5 The role of Venice 7
  1.6 Glass in Germany 8
  1.7 From art nouveau to modern glass design 10
  1.8 On the path to glass technology 10
  1.9 Otto Schott – founder of modern glass technology 11
  1.10 Glassmaking in the USA (rough outline) 14
  1.11 Glass all over the world 15

2 Glass, the material 16
  2.1 What is glass? 16
  2.2 General characteristics of the glassy state 17
  2.3 Broad classification of glass types 23
    2.3.1 Soda-lime glasses 23
    2.3.2 Lead glasses 24
    2.3.3 Borosilicate glasses 25
    2.3.4 Special glasses 25
  2.4 Raw materials for the manufacture of glass 26
    2.4.1 Soda ash 27
    2.4.2 Glauber's salt 27
    2.4.3 Potash 27
    2.4.4 Lime 28
    2.4.5 Alumina 28
    2.4.6 Lead oxides 29
    2.4.7 Barium oxide 29

Contents

	2.4.8	Boron compounds	29
	2.4.9	Coloring agents	29
	2.4.10	Opacifiers	30
	2.4.11	Glass recycling	30
	2.4.12	The batch	33

3 The glassmelt — 35
- 3.1 Melting furnaces and melting tanks — 35
  - 3.1.1 Pot melting — 35
  - 3.1.2 Tank melting — 37
  - 3.1.3 Tank construction — 39
  - 3.1.4 Materials for furnace construction — 39
- 3.2 Fuels — 41
  - 3.2.1 Gas — 41
  - 3.2.2 Fuel oil — 41
  - 3.2.3 Electricity — 42
  - 3.2.4 Heating — 42
- 3.3 The melting process — 44
  - 3.3.1 Primary melting — 44
  - 3.3.2 Refining — 45
  - 3.3.3 Conditioning — 45
  - 3.3.4 Refining in a tank furnace — 46
  - 3.3.5 Heat consumption in glass melting — 46
  - 3.3.6 Batch feeding — 47
  - 3.3.7 Melting defects — 48
  - 3.3.8 The sol-gel process — 49

4 Flat glass — 51
- 4.1 The production and use of common types of flat glass — 51
  - 4.1.1 Rolled (or cast) glass — 51
  - 4.1.2 Window and plate glass — 57
  - 4.1.3 Plate glass — 60
  - 4.1.4 Float glass — 60
- 4.2 Technical identification of soda-lime flat glasses — 63
- 4.3 Other types of flat glass — 63
  - 4.3.1 Antique glass — 64
  - 4.3.2 Flashed glass — 66

4.4	Processed flat glass	66
	4.4.1	Glasses with altered radiation, heat and sound transmission characteristics (solar, thermal and sound insulation)	66
	4.4.2	Non-reflective glasses	72
	4.4.3	Reflective flat glasses	73
	4.4.4	Other surface finishing techniques for flat glass	75
	4.4.5	Safety glass	76
	4.4.6	Fire-resisting glass	82

5			
Hollowware and glass tubing		84
5.1	The most important types of hollowware	84
5.2	The shaping of hollowware	85
	5.2.1	The mouth-blowing process	85
	5.2.2	Machine blowing	88
	5.2.3	Pressing	91
	5.2.4	Extrusion	92
	5.2.5	Spinning (centrifuging)	92
5.3	The drawing process for glass tubing	93
	5.3.1	Other tube drawing processes	93
5.4	Finishing of hollowware	95
	5.4.1	Torch blowing (lampworking)	95
	5.4.2	Industrial hollowware processing	96
	5.4.3	Insulating vessels	97
	5.4.4	Glass jewelry	98
5.5	Container glass	99
	5.5.1	Beverage bottles	100
	5.5.2	Bottling jars	102
5.6	Glass tableware	102
	5.6.1	Breakdown of tableware by glass type	103
5.7	Other hollowware	108
	5.7.1	Hollow structural glass	108
	5.7.2	Lighting glass	108
	5.7.3	Laboratory glass and medical hollowware	109
5.8	Finishing of hollowware	109
	5.8.1	Finishing in the hot state	110
	5.8.2	Finishing in the cold state – glass removing processes	115
	5.8.3	Surface coating processes	118

6	Special glasses and their uses		121
	6.1 Fused silica (fused quartz or quartz glass)		121
	6.2 Borosilicate glasses for industrial and laboratory use		123
		6.2.1 Laboratory equipment	123
		6.2.2 Glass process plant	125
	6.3 Pharmaceutical glass		127
	6.4 Glasses for electrotechnology and electronics		130
		6.4.1 Sealing glasses	130
		6.4.2 Glasses for television tubes	135
		6.4.3 Glasses for X-ray tubes, transmitting and image-intensifying tubes	136
		6.4.4 Glasses for soldering and passivation	137
		6.4.5 Sintered glass parts	139
		6.4.6 Glasses for high-voltage insulators	141
		6.4.7 Ultrasonic delay lines	141
		6.4.8 Electron conductive glasses	142
		6.4.9 Lamp glasses	143
	6.5 Electrode glasses		149
	6.6 Optical and ophthalmic glass		149
		6.6.1 Properties and classification of optical glasses	149
		6.6.2 Transmission of radiation; color filters	154
		6.6.3 Ophthalmic glass (spectacle glass)	157
		6.6.4 Special optical glasses for nuclear technology and radiation research	161
		6.6.5 The manufacture of optical glass	163
		6.6.6 Microspheres	164
	6.7 Glass fiber		168
		6.7.1 Insulating glass fibers	168
		6.7.2 Fiberglass textiles	169
		6.7.3 Glass fiber optics	173
	6.8 Glass-ceramics		183
	6.9 Porous glass and foam glass		186
	6.10 A glance into the future		188
7	Environmental protection in the glass melting process		190
	7.1 Glass melting		190
		7.1.1 Solid particle emissions	190
		7.1.2 Gaseous emissions	191
		7.1.3 Flue gas dust collection	191

	7.2	Waste disposal	192
8	Glass an an economic factor		194

Appendix	196
Glass museums	196
Explanation of physical symbols and units	196
Attenuation of radiation	197
Technical literature on glass	198

Index	199

Foreword

The manifold forms and uses of glass are becoming increasingly important in science, industry, and our personal lives. This constantly improving material interests a range of people extending beyond the relatively small number of glass experts. Naturally, questions arise as a result of this widespread interest. For this reason, we have heeded the publisher's suggestion to develop a glass primer which answers many questions and explains much of the terminology.

The bases for this Schott Guide to Glass were the lecture manuscript, 'Glass Science for Designers' by Prof. Dr.-Ing. Heinz Pfaender, and the Schott pamphlet, *Concepts of Technical Glass from A to Z*. The manuscript which evolved into this book was written by members of the Schott scientific staff. We thank all those involved in producing this reference work.

The *Schott Guide to Glass* will give experts, interested amateurs, and those who work with glass a glimpse into the diversity of this fascinating material.

Mainz, Germany, September 1995
The editor
Schott Glaswerke

Introduction

Glass is possibly the oldest man-made material, used without interruption since the beginning of recorded history. Unlike bronze or iron, however, it has not lent its name to any historical epoch. Still, the use of glass from hand-blown goblets to electronic components has grown with the rise of the industrial era and greatly affects present life. Glassmaking has always been one of the few truly integrated manufacturing processes where native minerals are transformed into an incredible variety of finished products within a single factory. Although glass production is an energy intensive process with fuel requirements which have changed the ecology and economy of large areas in past times, it uses common and virtually inexhaustible raw materials – in sharp contrast to the relatively scarce metallic ores.

The uses of glass are manifold and modern technologists are almost continuously proposing new applications. Glass is replacing less abundant and more expensive materials as witnessed by the growing importance of glass fibers for communication purposes. Nuclear technology would be unthinkable without radiation shielding apparatus made of glass, and there are many other areas where recently developed glasses are irreplaceable.

Perhaps the preceding millennia have not had a Glass Age because it is still to come. Scientists are already discussing metals and other materials which can be transformed into a glassy state, thereby opening up many new possibilities. Glass – a puzzling material? The *Schott Guide to Glass* contains the answers to most of the questions.

1
The history of glass

'Natural glass' is produced whenever glass-forming rocks melt under high temperature and then solidify quickly. This happens when volcanoes erupt, when lightning strikes into quartziferous sand or when meteorites hit the surface of the earth. During the stone age, humans used cutting tools made of natural glass of volcanic origin, known as obsidian and tektites.

Nobody knows exactly when glass was first made artificially by man. The oldest finds – greenish glass beads – date back to 3500 BC.

Glass originated in the Near East. The earliest finds are in Egypt and in Eastern Mesopotamia (Iraq). Glassmaking also developed independently in Mycenae (Greece), China, and North Tyrol.

Ancient glass manufacture is believed to be closely related to pottery making, which flourished in Upper Egypt about 8000 BC. While firing pottery, the coincidental presence of calciferous sand combined with soda and the overheating of the pottery kiln may have resulted in a colored glaze on the ceramics.

Not until 1500 BC was glass produced independently of ceramics and fashioned into separate items. Other theories consider glass to be a by-product of bronze smelting. Actually, glass and bronze often appear jointly in cultural history as there are close technical ties between the melting processes of these two oldest artistic materials.

1.1 GLASS IN EGYPT

For a long time, the development of glass was controlled by the status of melting techniques. Very few people mastered the art of glassmaking. The furnaces used in bronze smelting and

Fig. 1.1 Lotus goblet, Tutmosis III, approximately 1500 BC.

pottery making were too simple to melt a bubble-free, easily workable glass.

Around 3000 BC, Egyptian glassmakers systematically began making pieces of jewelry and small vessels. Starting in 1500 BC, the Egyptian glassmakers developed the sand core technique: they formed glass containers for unguents and oils around a solid core made of sand or ceramic. A craftsman dipped the core mold attached to a rod into molten glass, and the first usable hollowware was created. Constant rotation of the core in the molten glass caused the glass to adhere to the form. Rolling it on a flat stone slab smoothed the surface. Incised or shaped slabs allowed the surface to be decorated. Handles or carrying rings were then added (Fig. 1.1).

Available raw materials were used to melt glass. The addition of copper or cobalt compounds to the glass composition yielded blue tints. Glasses having a brownish appearance were also found. The clay tablet library of the Assyrian King Ashurbanipal (669–626 BC) contains cuneiform texts with glass formulas, the oldest of which reads approximately: 'Take 60 parts sand, 180 parts ashes of sea plants, 5 parts chalk – and you will get glass'. This glass composition contains essentially all the materials used today, although the proportions are somewhat inexact. The low amount of sand, however, leads to

A revolution in technology: the glassblowing pipe

the conclusion that achievable melting temperatures during the last millennium before Christ were still not very high, and it was only possible to make soft glass suitable for fashioning simple vessels and other wares. Through the course of the centuries, the art of glassmaking continued to spread. There were so many glassmaking operations in the Nile Valley from Alexandria to Luxor that one can speak of a glass industry of sorts. Similar developments occurred between the Tigris and Euphrates in Iraq, in Syria, on Cyprus and on Rhodes.

By 1000 BC, the glassmakers in the Eastern Mediterranean area and bordering regions were creating larger vessels and bowls through the development of new processes. For example, glass rods made of filaments of various colors were sliced and placed in molds, and glass was poured into the remaining spaces. Simple casting and pressing methods were also known. But all the technical possibilities were only good for the production of flat and deep bowls.

1.2 A REVOLUTION IN TECHNOLOGY: THE GLASSBLOWING PIPE

Around 200 BC Syrian craftsmen in the area between Sidon and Babylon made a decisive technical breakthrough with the discovery of the glassblowing pipe. This tool is made from an iron tube about 100 to 150 cm long with an opening about 1 cm in diameter. It has an insulated handle with a mouthpiece at one end and a button-like extension at the other end. The glassmaker uses it to get a gob of molten glass from the furnace and blows it into a hollow body. Since this time, the glassblower's pipe has only been slightly improved, despite technological progress. The blowing of glass with a pipe enabled not only simple, round vessels to be made, but also thin-walled, fine glasses in a large variety of shapes (Fig. 1.2). Using a wooden mold allowed the blowing of glass products in a standardized form and their duplication. Depressions fashioned into the molds, such as ribs, diamonds or nets, created decorations on the surface of the glass. The application of the glassblowing pipe was also the first step to making flat glass. In order to do this, glass was blown into large cylindrical bodies, then cut up, and 'ironed' flat while still in the hot state.

The well-developed trade relations among the peoples of the

Fig. 1.2 Roman glass objects.

Roman Empire, its highway and transport networks, and a Roman administration conducive to economic progress, were ideal prerequisites for the quick spread of the new invention and the art of glassmaking. In all parts of the Empire, from Mesopotamia to the British Isles, from the Iberian Peninsula to the Rhine, glassworks were founded. The craft experienced its first period of blossoming. Pliny the Elder (AD 23–79) described the composition and manufacture of glass in his encyclopedia, *Naturalis Historia*.

1.3 GLASS IN THE PERIOD OF THE ROMAN EMPIRE

In Alexandria around AD 100, the introduction of manganese oxide into the glass composition combined with improved furnaces resulted in the first successful attempt to make colorless glass. The capability to produce higher temperatures and to maintain better control over combustion atmosphere improved the quality of the glass due to more complete melting of the constituent materials.

The ostentation of the Roman emperors provided a further impetus to glassmaking. Artistically decorated luxury glasses with filigree, mosaic and engraved decors came into fashion. Glass was made into jewelry and used for imitations of

Fig. 1.3 Diatrate vase, fourth century.

precious stones. The antique art of glass coloring flourished (Fig. 1.3).

Roman glassworks usually settled in the vicinity of suitable deposits of sand. Labor imported from Alexandria played an important role. Until the Middle Ages soda was imported from Egypt and Syria. There were large numbers of glassworks in the Campagna region, as well as directly in Rome. Roman glassworks owners began identifying their products with their firms' logos as early as the first century and they sold them throughout the Empire. Roman glass specialities were shipped as far as China via the silk routes, although glass had already been developed there independently.

1.4 FROM LUXURY PRODUCT TO EVERYDAY ITEM

Numerous glassworks were in operation in the Syrian centers of Sidon and Tyre, in Egyptian Alexandria, in East Roman Byzantium, in Aquileia in Northern Italy, in the cities Amiens and Boulogne in northern France, and in the Germanic cities of Cologne and Trier (Fig. 1.4). For a long time, polished sheets of copper or silver served as mirrors. Then the Phoenicians created small glass mirrors with tin underlays. Since the flat glass they used did not have a smooth surface, glass mirrors

Fig. 1.4 Cabbage stem glass, fifteenth to sixteenth century.

posed no great threat to metal mirrors for hundreds of years. It was not until the thirteenth century in Germany, when the back side of a flat piece of glass was coated with a lead–antimony layer, that a quality mirror was successfully made in glass. This invention was later improved by the Venetian artisans, but the mirror format remained essentially unchanged.

Only after the development of the plate pouring process in 1688 in France under Louis XIV, could large-surface mirrors be manufactured. For this purpose glass was flattened on a pouring table with rollers, and after cooling, the surfaces were ground and polished smoothly and evenly. Thus plate glass, a flat glass of highest quality, appeared. By coating this glass with a low melting metal, it was converted into mirrors.

An ancient, longtime unfulfilled desire was to provide a transparent material for house windows. In ancient times parchment and oiled linen had to suffice as coverings for small window openings. Glazed windows were considered a great luxury until well into the Middle Ages. For centuries window panes were blown with a glassblowing pipe, cut open, and rolled flat. The window dimensions were very small, since the glassmaker could only handle a limited quantity of glass. The bull's-eye pane appeared in France during the fourteenth century. Its name is derived from the hub-like spiral, the bull's-eye, in the center. To make it, the glassblower first blew a glass ball which was then opened at the opposite side from where the glass was attached to the pipe. The glass was bent outward or spun flat. The finished panes had diameters up to 15 cm and were joined together with lead strips and pieced together into windows.

Among the oldest buildings with windows in Germany are the tenth century Tegernese Cloister and the eleventh century Augsburg Cathedral with its five prophet windows.

The peak period of stained glass began during the fifteenth century. Churches, palaces, town halls, guild halls, inns and residential homes had glass windows decorated with historical scenes or coats of arms. The spread of stained glass was presumably a direct result of the high windows in Gothic cathedral architecture. The use of colored glass moderated interior light levels and provided an appropriately moving ambience. Spectacle glasses were produced for the first time around 1250, simple microscopes and telescopes around 1600.

1.5 THE ROLE OF VENICE

In the Middle Ages, the old merchant metropolis of Venice gradually evolved into the center of Western glassmaking art. At one point, more than 8000 people were supposedly employed in the Venetian glass industry. The merchants of Venice dominated trade in the Mediterranean and between the fifteenth and seventeenth centuries glassmaking in the city reached its peak, not only in the actual glassmaking, but in the finishing work as well.

Venetian glassware designers were strongly influenced by many ideas from Islamic art. Syrian enamel painting was further developed here. The apex of Venetian glassmaking artistry was the creation of the purest crystal glass characterized by an inimitable glitter and absolute clarity. Pure quartz sand and potash made from sea plants were necessary for it (Fig. 1.5). Goblets with hollow stems and footed vases with reliefs of the lion's head of St. Mark characterized the peak of Venetian glass artistry. Bizarre winged glasses and cut cased glass mark the decline of the glassmaking art of the Renaissance in the seventeenth century. The cutting techniques and

Fig. 1.5 Gold-ruby bottle, probably Potsdam, late seventeenth century.

decorations exhibit a highly developed technology, but their effect is overdone and effete.

Glassmakers in northern Europe, primarily in the Netherlands and Germany, took up the Venetian tradition and spearheaded a transition to moderate design.

Venice jealously guarded its glass recipes, especially that for crystal glass. At one point, the glassmakers, who were housed on the island of Murano, faced death threats if they disclosed any formula. These master craftsmen held positions of high prestige, and not uncommonly attained ranks of nobility.

1.6 GLASS IN GERMANY

If we do not take into account the glassworks which were founded and operated in Germany throughout the Roman period and then disappeared, the beginning of a German glass industry goes back to the Middle Ages only. Expatriate Venetian glassmakers founded glassworks and also worked in Germany. They produced glasses in the Venetian style. German glassmakers settled in remote areas of the mountainous forests. An increasing quantity of glass was produced in the

Glass in Germany

Fig. 1.6 Model of glass melting furnace, according to George Agricola (1556).

Spessart Mountains, the Thuringian Forest, the Solling Mountains, the Black Forest, the Bavarian Forest, the Fichtel Mountains, the Bohemian Forest, the Erz Mountains, the Riesen Mountains and the Iser Mountains. Initially, a greenish sand- and potash-based glass was melted. Potash (potassium carbonate) was obtained from beech and oak wood. The tree trunks were burned, and the ashes were leached in containers, the so-called pots (hence, the name potash). The forests also served as sources of fuel for the glass furnaces. The finished product was called forest glass or *Waldglas*.

Once the surrounding forests were cut down, the glassworks (usually only quickly constructed wooden sheds for the furnace and for storage of finished glasses) were relocated (Fig. 1.6).

A good example of medieval German glass production is found in the Bavarian Forest. Its glass history is interesting because glass production remains the dominant industry in this region today, and the glass industry in other areas of Germany evolved similarly.

1.7 FROM ART NOUVEAU TO MODERN GLASS DESIGN

The beginning of the twentieth century saw the emergence of novel art nouveau glass shapes and glass decorations, not only in Europe but also in the USA. Among the artists designing glasses in the art nouveau style were Emile Gallé (1846–1904) in France, Louis Comfort Tiffany (1848–1933) in the USA, Josef Hoffmann (1870–1956) in Austria and Josef Maria Olbrich (1867–1908) and Karl Koepping (1848–1914) in Germany.

The Bauhaus influence (1919–1933) was also felt in glass design. The original Jenaer Glaswerk Schott & Gen.'s 'Sintrax' coffee machine was designed by Gerhard Marcks (1889–1981), who headed the Bauhaus Pottery. Wilhelm Wagenfeld (1900–1990), a Bauhaus disciple and master, from 1929 created a great many exemplary glass designs for a variety of German glass factories, including the renowned Jena tea set. Heinz Löffelhardt (1901–1979), one of the most outstanding glass designers in Germany, also co-operated closely with Schott-Zwiesel.

1.8 ON THE PATH TO GLASS TECHNOLOGY

The entire history of glass is characterized by the efforts of individuals who perfected and further developed production processes and products.

In 1676, English glassmakers developed lead crystal. The addition of lead oxide to the glass formula yielded a glass of high brilliance and pure ring. It was very suitable for deep cutting. This was not achievable on the continent for another hundred years. High purity lead glass was used as flint glass for optical purposes.

In 1679, Johann Kunckel (1630–1703), the director of the glassworks established near Potsdam by Friedrich Wilhelm of Prussia, the Great Elector, collected texts from his own experiments and those of others in his handbook, *Ars Virtraria Experimentalis*. This publication was recognized as the scientific basis of the German art of glassmaking until the nineteenth century.

Joseph Fraunhofer (1787–1826), the son of a glassmaker and a mirror maker himself by trade, got deeply involved in the technology of glassmaking. In Benediktbeuern he founded the first scientifically directed optical workshop. Here, he melted, formed and tested optical glasses, developed procedures for the calculation of lenses for telescopes and microscopes and explored new ways of building optical instruments.

The increasing exploitation of lignite and hard coal as fuels and the establishment of a soda industry released the glass industry from dependence on wood. The glassworks no longer had to be located in remote forested regions. The only requirement was an area with a useable transportation infrastructure.

The pot furnaces (in which ceramic pots with the raw materials were placed) which had been used since ancient times, were not sufficient for mass production. The invention of the tank furnace by Friedrich Siemens allowed continuous production and the use of machinery. Furnace technology was improved by the regenerative process, in which the exhaust heat from the melting furnace warms up burner gas and fresh air prior to their mixture, so that the fuel combustion is more efficient and higher melting temperatures can be achieved. Today's melting tanks have capacities of up to several hundreds of tons of molten glass.

Shortly before 1900, the American Michael Owens (1859–1923), invented the automatic bottle blowing machine which was introduced in Europe after the turn of the century. Somewhat later, processes for mechanical production of flat glass were available, without which the quickly rising demand for architectural glass could not have been met. For example, in 1851, 300 000 standardized glass panes were used as wall panels for the Crystal Palace built by Paxton for the Great Exhibition in London. This was one of the earliest examples of the use of glass as a structural material.

1.9 OTTO SCHOTT – FOUNDER OF MODERN GLASS TECHNOLOGY

Two German scientists laid the foundation for modern glass technology. Otto Schott (1851–1935), a chemist and technologist from a family of glassmakers, investigated the dependence of the physical qualities of glass on its composition by using scientific methods. In his father's basement laboratory, he studied the influence of many chemical elements on glass. In a manner of speaking, glass was rediscovered.

Fig. 1.7 Otto Schott.　　　　**Fig. 1.8** Ernst Abbe.

In 1879, Otto Schott contacted Ernst Abbe (1840–1905), who was a professor at the University of Jena and co-owner of the Carl Zeiss firm (Figs 1.7 and 1.8). Abbe needed suitable glasses for his high-quality optical instruments. The glasses had to be free of defects and of the highest purity. They also had to have consistent predeterminable optical properties. Lenses used so far suffered from the so-called secondary spectrum associated with previous glass types (Fig. 1.9).

After years of disappointing attempts, Otto Schott succeeded in producing an optical glass of the required quality on his 93rd trial melt. This was the breakthrough to the development of a variety of new optical glasses. He moved to Jena, and in 1884, along with Ernst Abbe, Carl Zeiss, and Zeiss' son Roderick, established the Glastechnisches Laboratorium Schott und Genossen which later became the Jenaer Glaswerk Schott & Gen. Schott, the 33 year old chemist, then devoted himself solely to glass research.

One of the first new products was a thermometer glass which hardly expanded when heated and therefore did not affect measurement precision. Other new types of glass and melting processes were conceived and investigated: technical (borosilicate) glasses capable of withstanding heat, pressure, and corrosion; optical glasses for microscopes and telescopes in

Translation:

                                            Witten, May 27, 1879

Professor Dr. Abbe
Jena

Dear Sir:

   I recently produced a glass, in which a considerable amount of lithium was introduced, and the specific gravity of which was relatively low. I suspect that such a glass will exhibit excellent optical properties, and I therefore wanted to inquire whether you or one of your colleagues might be willing to test it for refractive index and dispersion to determine whether my above supposition is correct. Your connections with Zeiss should make it easy for you to have the necessary grinding and polishing of the glass done. If you should be agreeable to my proposal, I would gladly send you several samples of the glass.

                                With deepest respect
                                and regards,

                                       Otto Schott

Address: Dr. Otto Schott
         Witten
         Westphalia

Fig. 1.9 The first letter from Otto to Ernst Abbe dated 27 May 1879.

small and large configurations; glasses for high-power camera lenses. Over the years, there were hardly any areas of industry which were not supplied with quality glasses from Jena. Cooking and baking utensils made of heat-resistant glass became common household items, making 'Jenaer Glas' known all over the world.

The Jena Glassworks soon had a worldwide reputation. It became evident that its founders also felt strongly about their social responsibilities. 'I do not intend to die a millionaire', declared Abbe. He transferred his wealth to the Carl Zeiss Foundation which he had founded. Otto Schott later added his share of the business to the foundation. The workers were assured of secure positions, shared the profits, and along with their families were cared for in cases of illness, disability, and death. The University of Jena received financial support. The firm was one of the few in the world to introduce the eight-hour work day in 1900.

In 1945, shortly after the end of World War II, the American army transferred key personnel from the Zeiss and Schott firms to West Germany. The present day headquarters of the Schott Group, the Schott Glaswerke, was set up in Mainz in 1952. The seat of the Zeiss Foundation was moved from Jena, East Germany to Heidenheim, West Germany. The Schott Group is the European leader in the development, manufacture, and sales of special glasses. Schott commands worldwide leadership in several areas, including optical glass. With approximately 18 000 employees worldwide, the group of companies manufactures more than 50 000 articles and has manufacturing and sales subsidiaries and agents in about 100 countries. The main industrial areas served are electrotechnology and electronics; optics and instrumentation; chemistry and pharmacy; information processing and data transmission; traffic technology; consumer products; and architecture.

1.10 GLASSMAKING IN THE USA (ROUGH OUTLINE)

- Mixed traditions of English and German glassmaking.
- Earliest efforts were colonial enterprises.
  1. Jamestown – failed enterprise (1600s).
  2. Annealing (1740s).
  3. Wistar – South Jersey Waldglas Industry (1780s).
- Glassmaking in the new nation – three major centers – nineteenth century.
  1. South Jersey – Waldglas, artistic ware.
  2. Pittsburgh, PA – Frontier glassmakers.
     Pittsburgh PA – Emphasis on energy and potash – forest products and NaCl utilized.
  3. Northern New York/New England – Glasshouses (Boston and Sulwich – Ressel glass). Technical innovations for artistic and consumer glass – forerunners of Corning.
- Glassmaking in the twentieth century.
  1. Growth of USA optical glass industry – Corning, Bausch and Lomb, American Optical, Pittsburgh Plate Glass.
  2. Industrial and consumer glasses – Owens-Illinois, PPG, Libbey, others.

1.11 GLASS ALL OVER THE WORLD

Basic ordinary glass quality is produced throughout the world. The most important raw materials and heating fuels are available nearly everywhere, and the necessary technology can be easily acquired. There is hardly any civilized country that does not produce glass. The building of manufacturing facilities for glass containers for food, drinks and household use usually marks the beginning of industrialization in the developing countries.

Thus, more and more nations are contributing to the history of the glass tradition which can be traced back over thousands of years. There are no indications that this trend will change soon, for the raw materials required to make glass are plentiful. In fact, glass has shown the potential to replace many materials which are becoming scarce.

2

Glass, the material

2.1 WHAT IS GLASS?

When considered as a material, glass is a collective term for an unlimited number of materials of different compositions in a glassy state. Glassy materials can also occur naturally. For example, obsidian, often found in volcanic areas, has a composition comparable to man-made glass. It consists of sand, and sodium and calcium compounds; and it was fashioned into knives, arrowheads, spearheads, and other weapons in ancient times. Natural glass in the form of obsidian was primarily used by peoples of the Eastern Mediterranean. It was in high demand in this region as an object of trade. The Aztecs in Mexico were also familiar with obsidian and created religious and household items from it.

When huge meteorites hit the surface of the earth, the released energy projected molten rock material into the atmosphere. This material fell back on the earth in the form of glassy lumps, so-called 'tektites'. Tektites are of bottle green to brownish-black color, the size between that of a walnut and a fist and have a glossy or scarred surface.

If lightning strikes sand, tubular vitrified crusts with a 0.5 to 2 mm wall thickness (fulgurites) are formed. They are 10 to 30 mm in diameter and several feet long.

Various chemical materials possess the capability for forming a vitreous structure. The most important among the inorganic materials are the oxides of silicon (Si), boron (B), germanium (Ge), phosphorus (P), and arsenic (As). When cooled quickly after melting, they solidify without crystallization, forming glass.

These glass formers exhibit the same behavior when mixed with other metallic constituents within certain system-dependent compositional limits. The addition of such 'glass modifying' components changes bonding relationships and structural groupings resulting in changes in the physical and chemical characteristics of the glasses. The glassy state is not limited to oxides however; it also appears when various sulfur and selenium compounds are rapidly cooled. Under some extreme circumstances, glass can be made from certain oxide-free metallic alloys. Many organic liquids also transform into the glassy state at low temperatures (such as glycerine at −90°C).

Until the eighteenth century, glass was exclusively made from sand, soda, potash, and lime. Coloring metallic oxides were also occasionally added. Today, about 60% of the 90 or so naturally occurring elements from hydrogen to uranium are used in the manufacture of glass. There are even some glass types (optical glasses, for example) that require almost 20 different ingredients. The possibilities for the applications of glass in science and technology are therefore extremely widespread.

Scientists have several answers to the question: 'What is glass?' One of the most popular is: 'Glass is an inorganic product of melting, which when cooled without crystallization, assumes a solid state', or: 'A frozen supercooled liquid is called glass'.

Actually, glass acts as an extremely viscous liquid which deforms very slowly under external force at normal temperatures. The naked eye cannot detect the deformation, but there are scientific procedures to calculate and measure it.

The following definition is more precise: Glass includes all materials which are structurally similar to a liquid. However, under ambient temperature they react to the impact of force with elastic deformation and therefore have to be considered as solids. In a more limited sense, the term 'glass' denotes all inorganic compounds which possess these basic qualities. At the same time, a clear line is drawn in comparison to plastics. Plastics are organic in nature and should never be designated as glass even if they are transparent.

2.2 GENERAL CHARACTERISTICS OF THE GLASSY STATE

Physically, all glasses are energetically unstable compared with a crystal of same composition. In general, when cooling a melted substance, crystallization should begin to occur when the temperature falls below the melting point (T_s). The reason this does not occur in glass lies in the fact that the molecular building blocks (SiO_4 tetrahedra in silicate glass, Fig. 2.1) are spatially cross-linked to one another. In order to form crystals, the linkage must first be broken so that crystal nuclei can form. This can only occur at lower temperatures. However, at those temperatures the viscosity of the melt impedes the restructuring of new molecules and thereby the growth of crystals. In general, the tendency to crystallize (glass specialists speak of 'devitrification') decreases with the increasing rate of cooling within the critical temperature range below T_s and with the number of components in the glass batch. Therefore, it is also influenced by the composition. The vitrification is only desirable in glass-ceramics (Glass-ceramics, Section 6.8).

The difference between a system (1), which crystallizes when brought below the melting point and the same system (2) which, as a result of preventing crystallization (by rapid cooling, for example), solidifies as glass, becomes especially clear if the volume is measured as the temperature falls. The result is shown schematically in Fig. 2.2; the temperature (T) is entered on the horizontal axis, the volume (V) on the vertical axis. As soon as the temperature falls to the melting point (T_s), System 1 jumps from A to B during which time it forms a crystalline mass. On the other hand, System 2, as a supercooled liquid, continues to shrink further to point C; and if the cooling is sufficiently slow, even further to point D. Here, the curve flattens out towards E or F, but it always remains above the base line B–G of System 1. Therefore at room temperature (T_R), the system does not reach the density of the crystallized system.

In the temperature range between point C and D, which is called the transformation temperature (T_g), the supercooled glass transforms from an elastic state to a viscous state typical for glass. The mobility of the structural elements is very low here, which is also demonstrated by the high viscosity of glass in this state (it is possible, for example, to determine the viscosity by the speed at which a horizontal glass rod supported at both ends is bent by a load placed at its center). It turns out that not only all inorganic glasses, but also all substances that exhibit the characteristics of System 1 in

General characteristics of the glassy state

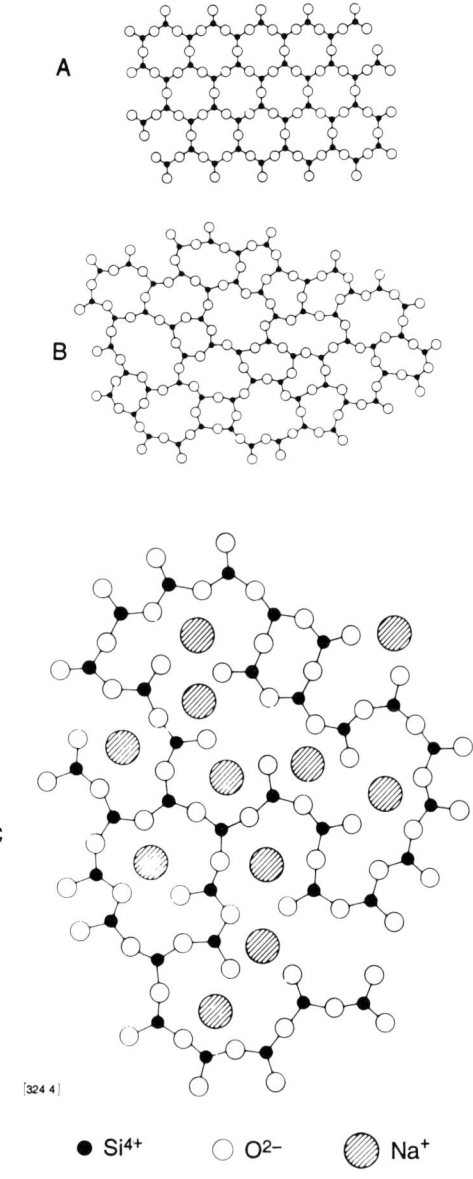

Fig. 2.1 Lattice of SiO$_4$ tetrahedra: (A) in crystal structure; (B) in fused silica structure; and (C) in sodium silicate glass structure (two dimensional display, the fourth oxygen bond of every tetrahedron is to be figured perpendicular to the perspective plane).

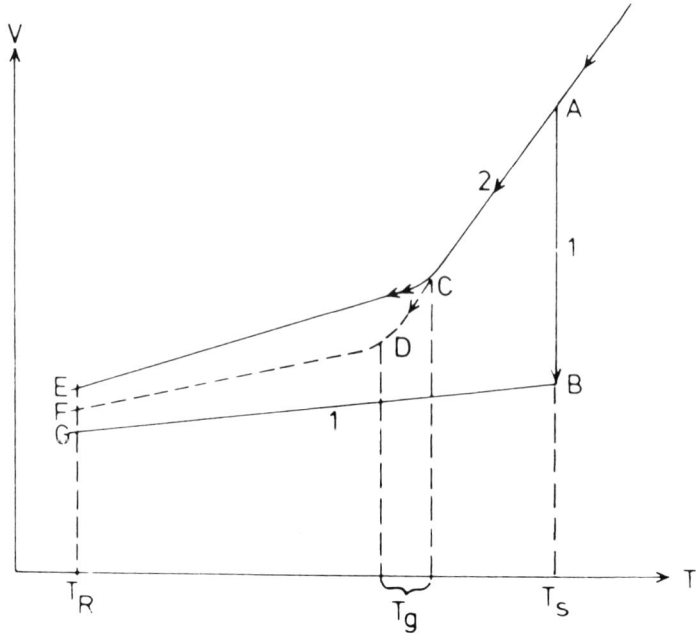

Fig. 2.2 Volume change during the cooling process of a melt forming: (1) a crystal structure; (2) a glass structure.

Fig. 2.2 have a viscosity value of $\eta \approx 10^{13}$ dPa s. (In the new SI units, the poise units have been replaced by dPa s (deci-pascal seconds), which are equal to 0.1 N s/m² (force unit N = newton, corresponding to a weight of 102 grams). The old unit, P, still appears frequently in glass literature (see Appendix).) In comparison, at 20°C, water has a viscosity of 10^{-2} dPa s, olive oil about 10^2 dPa s and honey about 10^4 dPa s.

The relationship between the viscosity of glass and the temperature (Fig. 2.3) is of basic importance in all facets of glass technology. It must be heated to a temperature at which $\eta \approx 10^2$ dPa s in order to achieve a homogeneous melt. Depending on the process used, glass can be worked and shaped at 10^3–10^8 dPa s. It the temperature range between viscosity values 10^4 and 10^8 dPa s is large, the glasses are called long glasses; if it is small, they are called short glasses. These differences are very important for processing because of the available glass working time.

General characteristics of the glassy state

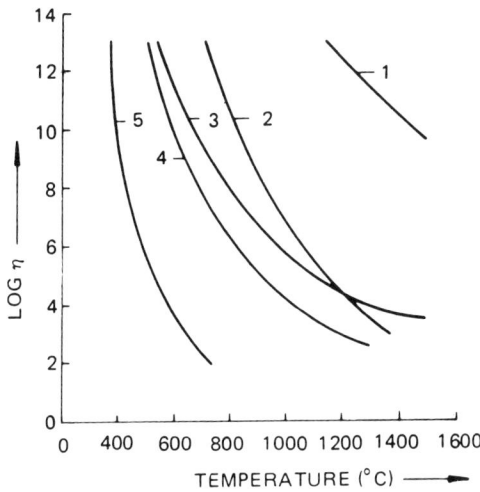

Fig. 2.3 Viscosity (η) versus temperature for various technical glasses: (1) fused silica; (2) alumina-silicate glass 8409 (see Table 6.3); (3) borosilicate glass *Duran* (Section 6.2); (4) soda-lime glass (Section 2.3.1); (5) lead borate solder glass.

In order to characterize the viscosity behavior of the different types of glass, several fixed viscosity values were established which describe the temperatures at which a particular glass can be processed accordingly. They are presented in the chart below.

For example, the fixed point temperatures of flat glass are illustrated in Table 4.1, and those of several technical glasses are listed in Table 6.3.

The transformation from the elastic to the viscous state occurs within the transformation range. In the same region, mechanical stresses are eliminated which may have developed due to too rapid cooling during processing (annealing). At

Table 2.1 Viscosity values for temperatures at which a glass can be processed

10^4 dPa s	Working point (V_A)
$10^{7.6}$ dPa s	Shortening point (E_W)
	Glass deforms under its own weight
10^{13} dPa s	Annealing point within the T_g area
$10^{14.4}$ dPa s	Strain point within the T_g area

10^{13} dPa s, 15 minutes are sufficient to accomplish this. At $10^{14.5}$ dPa s, the elimination of stress can take many hours. Stresses in the glass may be located and measured by the optical birefringence they cause.

The mechanical properties of glasses are also somewhat peculiar. Since the chemical bonding in the glass network and the energies required to break them are known, it is possible to derive theoretical tensile strengths. This results in values of the order of 10^4 N/mm^2. However, the technical strength is several hundred times less and is heavily dependent on the surface condition of the glass. For example, the maximum permissible continuous load for architectural glass is only set at 8 N/mm^2. If the glass is surface treated (e.g., fire-polishing, protective coating, and prestressing — see Safety glass, Section 4.4.5) to ensure that damage and microcracks are minimized, stability values of 5×10^2 N/mm^2 can be attained, which is still substantially below theoretical values. The reason for this is that internal defects and impurities are frozen within the glass during solidification. They limit the strength of the glass. The strength of thin, freshly drawn glass fibers for instance, is much higher because of the low number of flaws (Glass fiber, Section 6.7).

Experience with household glassware demonstrates that glass will break easily under quick temperature changes, especially if the temperature goes from hot to cold. Several properties lead to this behavior: poor heat conductivity, the relatively high thermal expansion of alkali-rich glasses, and limited tensile strength. When a glass is chilled (or quenched) after being heated to a point below its T_g temperature, only the outer layer begins to cool, and it seeks to decrease its volume. However, the core which is still hot stretches the outer layer and therefore subjects it to high tensile stress. If that stress exceeds the already reduced tensile strength of surface scratches, breakage occurs at that point and travels rapidly inward. On the other hand rapid heating is less dangerous since the outer layer comes under compressive stress, and the resistance of glass to compression is at least ten times that of its tensile strength.

Entirely different conditions are present when rapid cooling occurs above the T_g temperature. In this case compressive stress appears in the surface of the glass, increasing its tensile strength (Safety glass, Section 4.4.5).

Glass corrosion is defined as the behavior and eventual changes of glass subsequent to the attack of aggressive substances. There are two basic processes to be considered: first, the complete dissolution of the glass and second, the leaching. While the first is rare, the second process occurs more often. Leaching, for instance, causes sodium and potassium ions to be dissolved from the glass structure. The concentration of hydrogen ions (pH value) is changed by the implantation of hydrogenous ions in the glass, water molecules infiltrate the glass while alkali ions go into solution. A thin gel layer of low alkali and poor water content is formed, about 100 millionths of a millimeter thick (100 μm). If that layer grows thicker it becomes visible as dullness. Drinking glasses may show that appearance after several hundred cycles in a dishwasher. Cheap container glass (like mustard glasses), melted with a high addition of flux agents, show that condition very often. This is also the reason why glasses should not be stored in hygroscopic packaging material but rather in an aerated and dry environment.

2.3 BROAD CLASSIFICATION OF GLASS TYPES

The large number of glass types can be classified in several ways, for example, by chemical composition, use in the manufacture of glass products, or processing behavior. The most widely used classification is by chemical composition which leads to three chief groups: soda-lime glass, lead glass and borosilicate glass. Glasses in these categories account for at least 95% of all glass types. The remaining 5% are special glasses manufactured for the most part in very small quantities. Thousands of special glass types have been developed, and many of them have special applications. They are discussed separately in Chapter 6.

With very few exceptions, most glasses are silicate based glasses, the chief component of which is silicon dioxide (SiO_2).

2.3.1 Soda-lime glasses

By far the greatest number of industrially produced glasses belong to a group of glass types with very similar composition, collectively called the soda-lime glasses. As the name indicates, soda and lime play a major role along with the main component, sand.

A typical soda-lime glass is composed of 71–75% by weight sand (SiO_2), 12–16% soda (sodium oxide from the raw material soda ash or sodium carbonate), 10–15% lime (calcium oxide from the raw material limestone or calcium carbonate), and a low percentage of other materials for specific properties such as coloring. Sometimes magnesium replaces a portion of the calcium contained in the limestone, or potassium replaces the sodium in the soda. Even so, these glasses are similar to and may be classified as soda-lime glasses.

Soda-lime glass is primarily used for bottles, jars, everyday drinking glasses, and window glass.

The chemical and physical properties of soda-lime glass are the basis for its wide use. Among the most important is its light transmission which makes it suitable for use as flat glass in windows. An additional advantage is its smooth, nonporous surface which allows bottles and packaging glass made from it to be easily cleaned. Soda-lime glass containers filled with drinks and foods to not affect the taste of the contents, nor do they contain any harmful substances. Their resistance to aqueous solutions is sufficient to withstand repeated boiling (preserving jars) without creating negative surface changes.

Indeed, the relatively high alkali content of the glass lowers the melting point as opposed to pure SiO_2 glass but also causes an increase in the thermal expansion coefficient (α) by about 20 times from $\sim 0.5 \times 10^{-6}$/K to 9×10^{-6}/K (see Fused silica, Section 6.1 and Table 4.1). K (kelvin) corresponds to the old unit °C but is used for temperature differences.

With reference to the value α, all glasses are divided into two categories: glasses with α-values below 6×10^{-6}/K are called hard glasses; those with higher α-values are called soft glasses.

Owing to its high thermal expansion, the resistance of soda-lime glass to sudden temperature changes is comparatively poor (General characteristics, Section 2.2). Therefore, cautious handling is recommended, for example, when filling this glass with hot liquids.

2.3.2 Lead glasses

If lead oxide replaces much of the lime in the batch, the result is a glass type popularly known as lead crystal. Such glass is composed of 54–65% SiO_2, 18–38% lead oxide (PbO), 13–15%

soda (Na_2O) or potash (K_2O), and various other oxides. Glasses with less lead content (less than 18% PbO) are called crystal glass. Differing amounts of barium, zinc, and potassium oxides can be added to the composition to partially replace some of the lead oxide. To protect the consumer from misleading names, German legislation restricts the use of such terms as lead crystal, crystal glass, etc. to glass articles used as tableware and in the home (Glass tableware, Section 5.6).

Glasses containing lead exhibit a high refractive index and are especially suited for decorating by cutting. Their specific gravity is higher than that of soda-lime glass. In our daily lives, we usually see them as drinking glasses, vases, bowls, ashtrays, or as decorative items.

2.3.3 Borosilicate glasses

Silicate glasses containing boric oxide comprise the third group, borosilicate glass. These glasses have a higher percentage of SiO_2 (70–80%) than the previous two groups. The balance of the composition is as follows: 7–13% boric oxide (B_2O_3), 4–8% Na_2O and K_2O, and 2–7% aluminum oxide (Al_2O_3).

Glasses having such a composition show a high resistance to chemical corrosion and temperature change. For this reason, they are used in process plants in the chemical industry, in laboratories, as ampoules and vials in the pharmaceutical industry, and as bulbs for high-power lamps. But borosilicate glasses are also used in the home; baking and casserole dishes and other heat-resistant items can be made from it.

The family of borosilicate glasses is extremely broad, depending on how the boron compounds within the glassmelt interact with other metallic constituents. For this reason, most of these borosilicate glasses are classified as special glasses (Chapter 6).

2.3.4 Special glasses

Glasses used for special technical and scientific purposes form a mixed group. Their compositions differ greatly and involve numerous chemical elements. This group includes the optical

Fig. 2.4 Sand of high purity as a raw material for glass manufacturing.

glasses, glass for electrotechnology and electronics, and glass-ceramics. These are discussed in more detail in Chapter 6.

2.4 RAW MATERIALS FOR THE MANUFACTURE OF GLASS

Sand is the most important raw material for glass. Almost half the earth's solid surface consists of silicon dioxide (SiO_2), a main component of various sands and rocks. However, most sands do not have the purity necessary for the production of glass, as they have large amounts of coloring oxides, especially iron oxide. Sand containing as little as 0.1% Fe_2O_3 is useless for making such products as plate glass because it gives the glass a greenish tint. Natural sources for sands containing 0.01–0.03% Fe_2O_3 for the production of special technical glasses are very rare (Fig. 2.4).

Iron is an even greater problem in the quartz sand required for the melting of optical glass. The iron oxide content must be less than 0.001% (and sometimes even a small fraction of this), so low in fact that the content is expressed in parts per million (ppm), where 1 ppm = 10^{-4}%. The content of other coloring oxides, such as chromium, copper, nickel, cobalt, and other undesirable impurities in optical quartz sand must be substantially less than this. There are only a few deposits scattered over the earth's surface which meet these requirements. For these stringent demands, the crushed material is subjected to an additional chemical purification, using suitable acids at high temperatures. The grain size of sand ideally measures between 0.1 and 0.4 mm.

In order to lower the melting temperature of sand (over 1700°C) and to be able to use appropriate melting containers, a fluxing agent, usually sodium oxide, is required. This is normally added in the form of a carbonate (soda ash), or sometimes a nitrate or sulfate. However, alkali in this composition is seldom found naturally. Most of the alkali metals are bonded to halogens, mostly as sodium chloride (table salt). Thus glass manufacturing on an industrial scale could really only take place after it became technologically possible to transform alkali halides into oxide-bonded sodium. This was first made possible by the Le Blanc process, and later by the Solvay process for commercial soda (sodium carbonate) production.

2.4.1 Soda ash

Sodium carbonate (Na_2CO_3) is introduced to the glass batch as soda ash (an anhydrous, white powder). During the melting, the sodiun oxide becomes part of the glass; carbon dioxide is freed and escapes through the chimney.

2.4.2 Glauber's salt

Sodium sulfate (Na_2SO_4) (discovered in the seventh century by the doctor and chemist Johann Rudolph Glauber for medicinal use) can, in its anhydrous form, be mixed with pulverized coal and added to the batch instead of soda ash. The sulfurous acid is freed, and the sodium becomes part of the glass as in the case of soda ash.

2.4.3 Potash

Potassium carbonate (K_2CO_3) is a grainy, white powder formerly obtained by leaching wood ashes (mostly beech and oak) in large containers (pots). Today, it is commercially produced from potassium sulfate. The potash decomposes into potassium oxide, which goes into the glass, and carbon dioxide, which escapes into the atmosphere. In the absence of coloring metals, potash yields a pure, colorless glass.

Stabilizers to increase resistance, strength and hardness

There are a number of oxides derived from multivalent metals that can be added to the glass melt in order to give the glass physical and chemical properties that are important for its usability. Some substances can reinforce the structural network resulting in improved chemical resistance and mechanical properties. To this effect, the oxides of calcium (CaO), magnesium (MgO), aluminum (Al_2O_3) and zinc (ZnO), as well as boron trioxide (B_2O_3) play an important role. Replacing Na_2O with K_2O usually makes glass more chemically resistant.

2.4.4 Lime

Calcium carbonate ($CaCO_3$) is found naturally as limestone, marble, or chalk. At a temperature of about 1000°C, the carbon dioxide escapes from the lime. Only the calcium oxide (or calcined lime) remains, and it enters the glass structure. Lime is added to the batch to improve the hardness and chemical resistance of the glass. In flat glass, the lime is partially replaced by magnesium oxide which is found combined with lime in the raw material dolomite ($CaCO_3 + MgCO_3$). It lowers the melting temperature.

2.4.5 Alumina

Aluminum oxide (Al_2O_3) is usually added to the glass batch in the form of alkali containing feldspars abundant in many places (for example, $NaAlSi_3O_8$). The trivalent aluminum forms AlO_4-groupings in the glass which by inclusion of an alkali ion place themselves in the network of the SiO_4 tetrahedrons, thereby filling gaps. This leads to improved chemical resistance and increased viscosity in lower temperature ranges.

2.4.6 Lead oxides

The oxides, PbO (yellow lead) and Pb_3O_4 (red lead) are used to introduce lead into the glass. However, it is always present in glass as bivalent Pb^{2+}. Moderate additions of PbO into glass increase chemical resistance. High lead content lowers the melting temperature and results in decreased hardness, but increased refractive index of the glass, which is important for its 'brilliance' (Glass tableware, Section 5.6).

2.4.7 Barium oxide

Barium oxide derived from $BaCO_3$ (witherite) is primarily used in optical glass and crystal glass instead of lime or red lead. Glass containing barium is not quite as heavy as lead crystal, but achieves similar brilliance due to its high refractive index.

2.4.8 Boron compounds

Boron trioxide (B_2O_3, the anhydride of boric acid, H_3BO_3), which is so important in special glasses, is found naturally in very few places. Sodium and calcium borates occur much more frequently. Usually, these compounds must be chemically converted into pure boric acid, especially for optical glass.

2.4.9 Coloring agents

Only pure chemicals are used as glass coloring agents. Different colors can be obtained by adding oxides of the so-called transition elements (copper, chromium, manganese, iron, cobalt, nickel, vanadium, titanium) or of the rare earths (primarily neodymium and praseodymium) to suitable base glassmelts. The colors yielded by these metallic ions are listed in Table 2.2.

Intensive yellow, orange and red colors are produced by the precipitation of precious metal colloids as well as of selenium, cadmium sulfide, and cadmium selenide during the cooling of the melt. Secondary heat treatments are also used to obtain these colors (struck glasses, the best-known of which is gold-ruby glass). Brown and gray colors are obtained with combinations of oxides of Mn, Fe, Ni, and Co, which in high

Table 2.2 Coloring agents added to base glassmelts

Copper	(Cu^{2+})	light blue
Chromium	(Cr^{3+})	green
	(Cr^{6+})	yellow
Manganese	(Mn^{3+})	violet
Iron	(Fe^{3+})	yellowish-brown; see also 'carbon yellow' coloring, (Container glass, Section 5.5)
	(Fe^{2+})	bluish-green
Cobalt	(CO^{2+})	intense blue, in borate glasses, pink
	(CO^{3+})	green
Nickel	(Ni^{2+})	grayish-brown, yellow, green, blue to violet, according to the glass matrix
Vanadium	(V^{3+})	green in silicate glass; brown in borate glass
Titanium	(Ti^{3+})	violet (melting under reducing conditions)
Neodymium	(Nd^{3+})	reddish-violet
Praseodymium	(Pr^{3+})	light green

concentrations can also yield black. Undesirable color tinges can be eliminated where necessary by using oxygen releasing substances such as refining agents (The melting process, Section 3.3).

At 400–600°C, the surface of colorless glass can also be stained from yellow to reddish-brown. Silver stain is especially popular (Finishing, Section 5.8).

2.4.10 Opacifiers

By mixing fluorine containing materials, such a fluorspar (CaF_2) or nowadays preferably phosphates, small crystalline particles form in the glass, rendering the glass cloudy and opaque. Such glasses are used in opal tableware, in opaque or milk glass for architecture, and in flashed opal glass globes for the lighting industry (see Chapters 4 and 5).

2.4.11 Glass recycling

Glass cullet from new melt

Although glass cullet is not technically a raw material, it must be added to the melt. Each glass factory saves its cullet, be it in

Fig. 2.5 A 30–40% content of good quality glass cullets in a melt optimizes the tank efficiency.

the form of pieces cut from flat glass or rejects or breakage from hollow glass (Fig. 2.5). If the cullet stock is insufficient it may be 'produced' or purchased. Cullet acts as a fluxing agent and accelerates the melting of the sand. This conserves both energy and raw materials.

Glass cullet from scrap glass

In order to reduce the discharge on landfills and, as previously described, to lower the energy and raw material consumption, scrap glass recycling is of growing importance. In Germany and many other countries where environmental consciousness is well advanced, scrap glass is widely collected in waste-glass containers and special trucks and then brought to recycling plants (Fig. 2.6). Here the scrap glass is either manually or automatically sorted into white and colored glass according to its further use. Foreign, non-glassy objects are also sorted out. After the glass is crushed into smaller pieces, additional

32   *Glass, the material*

Fig. 2.6 Container for recyclable used glass.

Fig. 2.7 A batch.

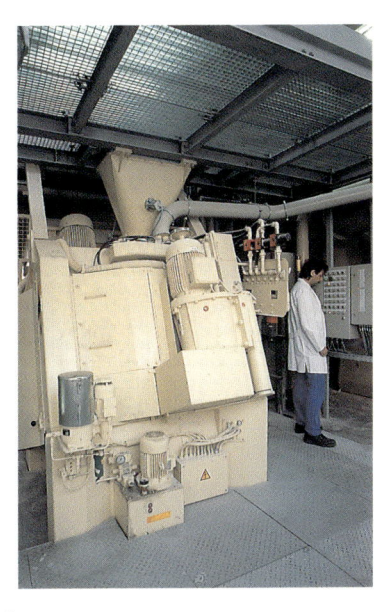

Fig. 2.8 A batch house.

cleansing from foreign material is done using magnets or blowers, for instance. Some melting plants are equipped with processing lines to clean and prepare the waste glass for its reuse in new glass melts.

Every ton of waste glass reused in a new melt replaces 1.2 tons of raw material and saves 100 kg of fuel. It is conceivable that glass can be almost 100% recycled, provided that scrap glass is selectively sorted by colors and processed with the most advanced technology available. Presently the share of glass cullet in a new melt is about 30% for white and brown glass and over 90% for green glass.

2.4.12 The batch

The mixture of the individual raw materials combined in certain amounts to yield the desired glass type is called the glass charge. When the glass charge is carefully put together, mixed, and ready for melting, it is called a batch (Fig. 2.7).

All raw materials are intensively inspected and tested for quality when they enter the glass plant. Based on the results, the batch house (Fig. 2.8) calculates any correction factors in order to compensate for composition variations of the raw materials. With the aid of electronic programming, batch composition can be fully automated in modern plants. To prevent the mixed ingredients from segregating prior to melting, batches are prepared with a water content 2–4%.

3
The glassmelt

Melting is the central phase in the production of glass. The individual raw materials combine at high temperatures to form molten glass. The quality of the batch material, the type of heating energy, and the type of melting process used are determined by the glass type to be melted and the product to be made. Removal of the glass from the furnace for further processing and the cooling cycle are stages subsequent to the melting process.

3.1 MELTING FURNACES AND MELTING TANKS

Originally, only crucible and pot furnaces were used. They could hold one or more containers to melt the batch material. From these, continuous melting tanks evolved as an integral part of the furnace. Both types are still frequently used side by side.

3.1.1 Pot melting

A pot furnace consists of refractory brick for the inner walls, silica brick for the vaulted roof (crown), and insulating brick for the external walls. Basically a pot furnace consists of a lower section to preheat the fuel gas and the upper furnace which takes the pots and serves the melting chamber (Fig. 3.1).

Today, pot furnaces are only used to manufacture mouth-blown glass products and special glasses. There is room for six to twelve pots in which different glass types can be melted (Fig. 3.2).

A pot is made of refractory clay (sometimes mixed with sillimanite 'grog'). It has to be heated slowly to 1000°C before it can be used to melt glass. It resembles an upside-down,

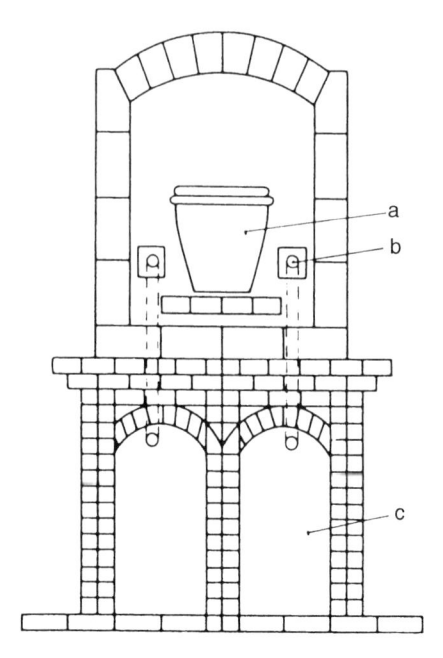

Fig. 3.1 Structure of a pot furnace (schematic) with: (a) melting pot; (b) burner ports; (c) regenerative chambers for heat recovery process.

rimless hat, open at the top. In the USA, covered pots are generally used. They have an opening at the top where the batch can be added and glass removed. Formerly, glass factories manufactured their own pots in their production facility. Today, they are chiefly supplied by companies specializing in ceramic pot production. The capacity of a pot can be as great as 2000 kg, but it is usually between 100 and 500 kg. The working lifespan of a pot, if operated continuously averages 2–3 months (Fig. 3.3).

Pot furnaces are heated 24 hours a day, but their temperatures vary. The batch is loaded into the pot and melted in the afternoon. The temperatures are increased overnight to refine the melt so that the glass can be processed beginning early next morning. During melting, the temperature climbs to 1300–1600°C, depending on the glass type. During removal and processing of the glass, the temperature in the furnace is 900–1200°C. Pot melting is not suitable for continuous production, as the molten glass normally lasts for one shift only.

Melting furnaces and melting tanks 37

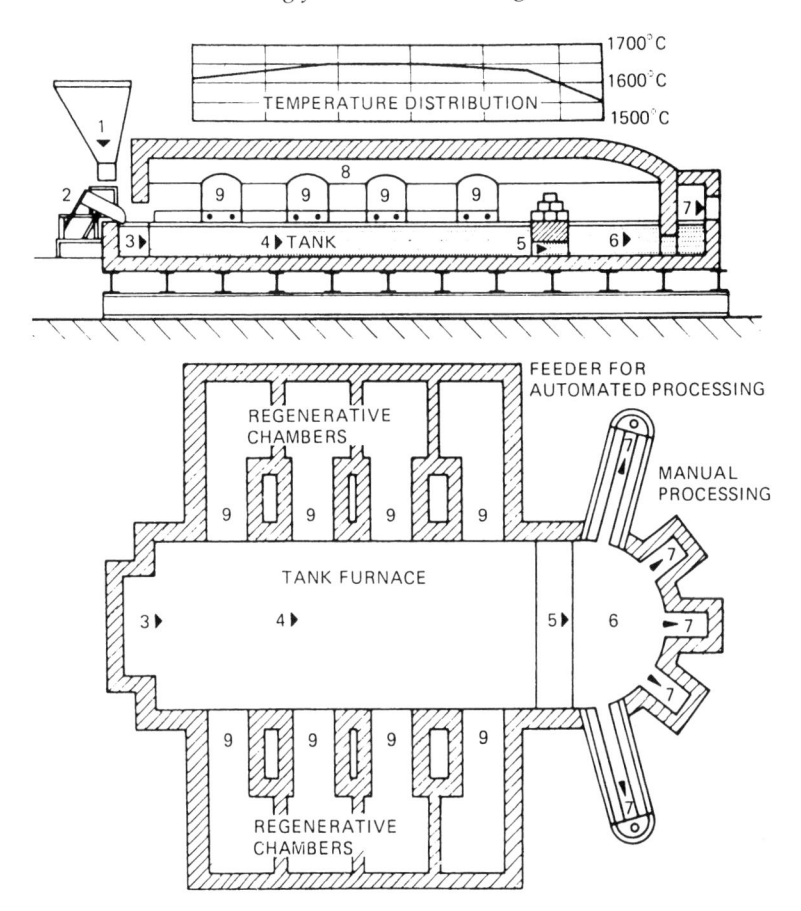

Fig. 3.2 Floor plan and cross section of a melting tank: (1) glass batch container; (2) batch feeder; (3) batch feeding compartment ('doghouse'); (4) melting and refining tank; (5) tank throat ('doghole'); (6) forehearth; (7) glass feeder for automated and manual processing; (8) roof (or crown) of the melting tank; (9) burner ports (in pairs) for combustion gas and flue gas, respectively, on opposite sides of the tank.

3.1.2 Tank melting

When large amounts of glass have to be processed, and above all, for automated glass production, tank furnaces are used. They consist of lower and upper sections as well as chambers

Fig. 3.3 Cast of an optical glass block from a pot melt.

in which the combustion air is preheated. The lower section is the actual melting container lined with a high quality refractory material. The size and construction of tank furnaces depends on the glass product manufactured. There are day tanks and continuous tanks (Fig. 3.4).

Day tanks

Day tanks are further developed from the traditional pot furnaces for small capacities (such as 10 tons per day). They are refilled with batch daily. The melting is done at night, and the glass goes into production the next day – indicative of their close technical relationship with pot furnaces. They allow a change in glass type to be melted on short notice. Day tanks are primarily used for colored glass, crystal glass and soft special glasses.

Continuous tanks

The development of continuous tanks was a prerequisite for industrial mass production and the rise of large glass companies. They are exclusively used in the manufacture of flat glass, container glass, and certain mass produced optical glass types. Continuous tanks are 10–40 meters long and 3–6 meters wide. Their output lies between 100 and 400 tons per day, depending on the product. This requires continuous feeding of batch material (Fig. 3.5) and continuous processing of the molten glass. The processing teams at these tanks almost

Fig. 3.4 Roof (or crown) of a melting tank.

always work in three shifts with fully automated production. Except for blast furnaces used in the iron and steel industry, glassmelting tanks are the largest industrial furnaces. The giants among them, used for float glass production, have a length of 100 meters and a width of 13 meters. The melting tank itself can contain up to 2500 tons of molten glass.

3.1.3 Tank construction

Continuous tanks differ primarily in size and heating systems. The most common types are regenerative side-burner tanks, with preheating of fuel and combustion air; regeneratively heated U-tanks for small output, which require relatively little space; and unit melters or long tanks, a special design for bottle manufacturing with horizontally opposed burner systems saving space, but consuming larger amounts of energy.

3.1.4 Materials for furnace construction

The refractory materials utilized in the glass furnace construction, regardless of the type of furnace, are used up over time.

Fig. 3.5 Automated batch feeding.

Their suitability depends on their working lifespan, on the least possible contamination of the melt by dissolution of the material, and on the cost. Major developments in furnace construction material in the last four decades have led to increased melting temperatures, improved melting efficiency, and better glass quality. The most critical furnace components are currently made of cast fused corundum-zirconium oxide refractory. The life of a tank furnace, called the furnace journey, is currently about eight years. The normal lifespan of a pot furnace is approximately five years. Within this time period it is possible to repair the furnace; these can be either hot (without interrupting the heating operation) or cold repairs (the furnace is shut down). Contamination of the melt by dissolved lining material is particularly critical in optical glasses. To avoid this, tanks and pots are lined with platinum. Platinum and its alloys (e.g. with rhodium) are insoluble in molten glass, have a high melting point and maintain consistently the same properties in a constant composition.

In many cases, other components of the furnace that are in contact with the melt, e.g., feeder, doghouse-lining, bubble-valves, electrodes, stirrer coatings, etc., are made of platinum, as well.

3.2 FUELS

As mentioned in Chapter 1, 'The history of glass', wood was originally the only available fuel for glass melting. It was gradually replaced by coal, beginning in the seventeenth century, but coal (both lignite and hard coal) was not in general use until the beginning of the nineteenth century.

3.2.1 Gas

Like wood, coal was originally used to fire glass furnaces directly. However in 1860, Wilhelm Siemens developed indirect heating by burning gas which was obtained by degassing coal in a gas generator. This method was quickly adopted by the glass industry and generator gas firing was not replaced by more modern methods until after 1950. Then plants began to switch more and more from producing their own gas to purchasing it from a gas company. The quickly growing availability of natural gas at the end of the 1960s accelerated this trend. The advantages of modern gas firing are high purity, ease of control, and the elimination of storage costs.

Where glass plants are not located close to the gas pipeline, liquid gases (propane or butane) are used to heat small melting and processing facilities.

3.2.2 Fuel oil

Fuel oil quickly became popular in the German glass industry after 1950. There are light and heavy heating oils with varying heating values in use. One advantage of fuel oil over gas is the brighter burning flame that can more easily be adjusted;

disadvantages are the high sulfur content of some fuel oils, the tendency to coke, and the on-site storage in tanks.

3.2.3 Electricity

The use of electricity to melt glass goes back to around 1900. Several methods are being used: radiation of the glass melt surface with heating elements, induction heating with medium frequency AC, and resistance heating in the tank lining. Heating the glassmelt with submerged electrodes is more effective and gaining increased popularity. In countries where electricity costs are low (Scandinavia), electric glassmelting enjoys widespread use. In fully electrically heated tanks, a heating efficiency (expressed simply as the ratio of effectively used heat to supplied heat) of over 70% can be achieved; with flame heating (gas or fuel oil), 30% is seldom exceeded. In addition, electro-melting will not harm the environment, adds no impurities to the glass, and requires limited investment and repair costs. Electric melting of glass is by far the most economical method in smaller tanks which use less primary energy than comparable flame heated tanks. Electricity prices will determine whether this method will continue to spread (Fig. 3.6).

3.2.4 Heating

When the furnaces are heated with gas or fuel oil, either both combustion gases (gas and air) or the air alone is preheated. The preheating has two effects: first, the fuel consumption is reduced by 50–70%, second, this is very often the only way to achieve the required furnace temperature of up to 1650°C.

Regenerative heating

The principle of regenerative heating consists of directing the combustion air through preheating chambers, which are connected in pairs to the overhead structure, before it enters the furnace. Preheating chambers are filled with an open brickwork offering a large contact surface for air and gas. They work in alternating cycles. The hot flue gases of the melt heat up one half of the preheating chambers. After half an hour, the flow of

Fuels

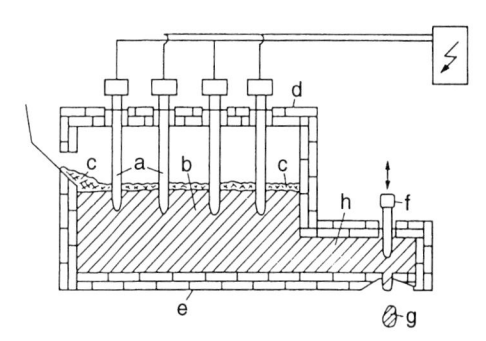

Fig. 3.6 Structure of an electrically heated tank (schematic): (a) platinum electrodes; (b) glass melt; (c) batch; (d) crown; (e) tank floor; (f) plunger; (g) metered glass gob; (h) forehearth.

the cold combustion gas is reversed and directed through the preheated brickwork while the flue gases heat up the other half of the preheating chambers. After another half hour, the flow of the cold combustion gas is reversed and directed through the preheated brickwork while the flue gases heat up the cooled down other half of the preheating chambers. This method can be used for fuel oil and gas heating.

Recuperative heating

This heating method does not alternate cold air and hot flue gas flow within the furnace system. The flow direction of the combustion gases (gas or heating oil) and the flue gases is always the same. The exchange of heat is accomplished by means of a thin separator wall between the cold air and hot flue gases which facilitates heat exchange.

Flue gases

The flue gases escaping from the melt are filtered through appropriate equipment. The use of energies low in contaminating components, such as natural gas, and the extensive use of recyclable waste glass, helps lower the amount of flue gas emissions. The emission of carbon dioxide (CO_2), heavily influencing our climate, has been reduced to one-third per ton of glass over the past 20 years. Filter dust, retained from the flue gases, is now reused and added to the glass batch.

Fig. 3.7 Glass pot melt at approximately 1200°C.

3.3 THE MELTING PROCESS

The melting process is divided into several phases which require close supervision and control. This holds true for the traditional pot furnaces as well as the most modern tank furnaces.

3.3.1 Primary melting

Owing to the poor thermal conductivity of the batch, the temperature spreading in a newly added batch is so slow that there is time for different physical and chemical processes and reactions to occur between the various components of the batch. Some of the raw materials decompose under the effect of the heat, gases trapped in the raw material escape, and the moisture in the batch evaporates. The raw materials slowly melt between 1000 and 1200°C (Fig. 3.7). First, the sand begins to dissolve under the influence of the fluxing agents. The silicic acid from the sand combines with the sodium oxide in the soda ash (or the potassium oxide, depending on the fluxing agent) and with other vitrifying batch materials. At the same time, large amounts of gases escape through the decomposition of the hydrates, carbonates, nitrates and sulfates, generating water, carbon dioxide, nitrogen, oxygen and sulfur dioxide. For example, one liter of a soda-lime glass batch sets free about 1440 liters of gas at 1000°C, 70% of which is carbon dioxide

(CO_2). In reality, the glassmelting process is, of course, much more complicated. The glassmelt finally becomes transparent, and the melting phase is completed. The volume of the original tank or pot load has decreased to about a third, because the small, empty spaces between the grains of batch material have disappeared and all gases have escaped. In pot melting, more batch material must be added in order to get the required amount of glassmelt. In tank furnaces, batch material is usually added continuously by automatic batch chargers.

3.3.2 Refining

While the batch is melting, the long process of homogenization begins. Homogenization involves the complete dissolution and even distribution of all components (especially important for the elimination of striae), as well as the refining process which eliminates all bubbles from the melt. In order to achieve the most homogeneous melt, free of bubbles, a thorough mixing and degassing of the glass is necessary.

Chemical refining agents are most frequently used to accomplish this based on the fact that the added compound sets free gas (for example, arsenic pentoxide (As_2O_5) decomposes into arsenic trioxide (As_2O_3) and oxygen at approximately 1250°C). Sodium sulfate (Glauber's salt) has a similar effect, in that it releases sulfur dioxide and oxygen at approximately 1200°C. It is preferred as a refining agent for mass-produced glass because it is the least expensive. The oxygen bubbles absorb the other dissolved gases and bubbles, thereby growing bigger so that they rise to the surface faster.

Refining can also be facilitated and stabilized by the bubbling process. Steam, oxygen, nitrogen or air is forced through openings in the bottom of the tank. By further increasing the temperature the glass becomes less viscous and the gas bubbles can rise more easily to the surface. This is known as plaining. Stirring mechanisms are used in the melting of optical glass in order to obtain the high degree of homogeneity required.

3.3.3 Conditioning

A conditioning phase at lower temperatures follows the primary melting and refining stages. During this process, all remaining soluble bubbles are reabsorbed into the melt. The dissolution of residual bubbles is an important part of refining. At the same time, the melt cools slowly to a working temperature between 900 and 1200°C.

In pot melting, these steps occur in sequence. In tank furnaces, which are in continuous operation, the melting phases occur in different locations within the tank. In other words, the batch is fed at one end of the tank and flows through different zones in the tank where primary melting, refining, and conditioning occur, until the working temperature is reached at the other end of the tank. The refining process in a tank is actually more delicate. Because of the importance of tank melting for commercial glass manufacturing, it is described separately.

3.3.4 Refining in a tank furnace

Glass does not flow through the tank in a straight line from the hot end batch feeder to the cold end where the glass reaches the working temperature for processing. It is diverted following thermal currents. The batch pile, or the cold mixture of raw materials, is not only melted at the surface, but also from the underside by the molten glass bath. Relatively cold, bubbly glass forms below the bottom layer of batch material and sinks to the bottom of the tank. Were it not to reach the surface again due to thermal currents, it would not be refined, since refining occurs in tank furnaces primarily at the surface of the melt, where bubbles need to rise only a short distance to escape.

Furthermore, only there are the temperatures high enough to cause bubble growth sufficient for refining. If thermal currents flow too fast, they inhibit refining by bringing the glass to the conditioning zone too soon. Guiding walls are built into the inner tank structure to create ideal glass flow paths.

3.3.5 Heat consumption in glass melting

The theoretical heat requirement is the amount of heat energy necessary to melt the glass and achieve the required melt temperature. It is made up of the melting heat (the heat required to transform the batch into glass), and the retained heat in the glass and gases formed up until the glass reaches

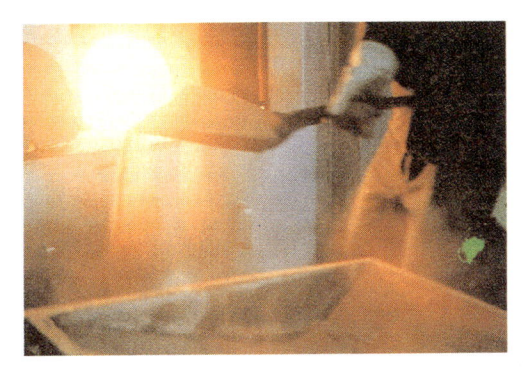

Fig. 3.8 Batch feeding of a pot furnace.

the end temperature. To evaluate the efficiency of tank furnaces, data on idle furnace conditions, daily throughput, glass take-out ratio, daily heat energy consumption and heat transfer efficiency are required. The idle furnace value is the heat required to maintain furnace temperature with no glass take-out. Heat transfer efficiency is the ratio of available heat (equal to melting capacity multiplied by theoretical heat consumption) to actual heat consumption.

3.3.6 Batch feeding

In pot furnaces the batch material is added with a shovel (Fig. 3.8). When the furnace is filled with molten, refined and conditioned glass, the glass is processed. Only then is new batch material added. This is a 24-hour cycle. It is done differently in most modern tank furnaces which are usually constructed with a batch feeding compartment (the 'doghouse'). The molten glass surface in the doghouse is covered with batch material and the floating batch is pushed away. The advantages of feeding the furnace through the doghouse lie in the reduced amount of dust development and deposit and consequently longer furnace life, less radiant heat loss through the feeder opening, ease of operation, and direct connection to the batch silo.

Automatic ram-feeding equipment was developed in the 1980s. With this method, batch feeding machines cover a large part of the melt surface with a continuous layer of batch

Fig. 3.9 Glass sampling from a tank furnace.

material, 3–8 cm thick and almost reaching to the tank side walls. This has a beneficial effect on throughput capacity. Considerable increases in throughput can be achieved, maintaining the same melting temperature and glass quality. The most important requirement for the mechanical batch feeder in tank furnaces is to maintain a constant glass level. The latter is automatically measured by mechanical, electrical and optical sensors that control the batch feeding rate.

3.3.7 Melting defects

There are a number of defects caused by the melt and visible in the finished product that impair its usability unless they are deliberately produced for artistic effects. These defects can be tested for in the melt (Fig. 3.9).

Stones

These are nontransparent grains of various sizes. Devitrification stones, as they are called, appear when molten glass crystallizes. Stones can also be small particles of the refractory material which have broken off the tank or pot walls. Batch and melt stones result from raw material grain sizes being too large, or from an insufficient melting temperature.

Striae (cord)

These are streaks of inhomogeneous, transparent glass in glass. Optically, striae are zones of glass with a different

Fig. 3.10 Striae in glass.

refractive index, producing distortion of light transmission visible in polished articles. Very intense striae are known as cords and have the appearance of ropes or strands within the bulk of a piece of glass (Fig. 3.10). During the processing of glass, striae and strings can also occur on the surface of the glass product.

Bubbles

These are mostly caused by insufficient refining of the melt. Large bubbles are referred to as blisters and small bubbles as seeds. Longitudinally stretched bubbles are often called airlines.

Discolorations

These are caused by impurities in the raw materials or insufficient decolorizing of the melt.

3.3.8 The sol-gel process

A completely different way of making glass without a melting process is the so-called sol-gel process. To do this, organic compounds, such as alcoholates of silicon, sodium or calcium, react with water in a hydrolysis process and split into water and alcohol. This process also creates a structure in which the metallic atoms are bonded to oxygen atoms in a non-crystalline irregular network, forming a gel. A low temperature treatment transforms the gel into an inorganic glass.

The sol-gel process produces high purity glasses, free of foreign material and with a high homogeneity; however, glass objects of larger dimensions require a long processing time and so at this stage of development, a larger scale application of this process is only appropriate for thin film coatings. Typical applications are heat or infrared reflecting coatings on windows or coatings on rear-view mirrors to reduce glare and reflection.

4

Flat glass

The term flat glass pertains to all glasses produced in a flat form, regardless of the method of manufacture. Unlike hollow-ware, it was not until the Middle Ages that glassblowers were finally successful in making flat sheets of glass for windows and further processing. Mechanical production processes were developed much later, effectively after 1920 (apart from some tentative earlier beginnings). Today manual production of flat glasses is the exception rather than the rule.

Over one-third of the flat glass produced in Germany is not used in its original form, but converted into other products, such as automotive safety glass and mirrors.

By far the greatest amount of flat glass produced consists of soda-lime glass. All manufacturers use basically the same formula. Average ingredients are: 72% SiO_2, 14% Na_2O (+K_2O), 9% CaO, 3–4% MgO, and 1% Al_2O_3. The iron content of the raw materials gives thick soda-lime flat glasses a green tint. Different glass tints for special purposes are also produced by adding coloring oxides to the same base glass mix. Borosilicate sheet glasses and 'white' special mirror glasses, on the other hand, are virtually colorless (Other types of flat glass, Section 4.3). Their iron content ($\leq$ 0.03%) is 5–10 times less than that of common soda-lime glass.

4.1 THE PRODUCTION AND USE OF COMMON TYPES OF FLAT GLASS

4.1.1 Rolled (or cast) glass

Rolled glass is a poured and rolled flat glass which in not clearly transparent.

Production of rolled glass

The molten glass flows from the melting tank over a refractory barrier called the weir onto the machine slab. From there, the glass flow enters between two water-cooled rollers. Their distance from each other determines the thickness of the glass plate. A refractory 'gate' controls the amount of the glass flow. If wire-reinforced glass is made, the wire feeding equipment is located in front of the rollers. The wire netting is pressed into the hot glass by means of a locating roller. If the rolled glass needs to be textured, that is, if the surface needs to have a certain design, patterned rollers are used with the desired surface relief. They work like a continuous embossing stamp. The application of the surface design precedes the annealing process. The glowing ribbon of glass is conveyed over rollers into a tunnel-like annealing furnace or lehr (Fig. 4.1).

Inside, the cast glass is first reheated to 600–800°C. On the way through the annealing tunnel, the temperature decreases slowly in an exactly calculated manner, so that no internal stresses can build up in the glass. At the cold end, the finished product, just lukewarm, is ready for use. It is cut to finished and standard sizes and packed in shipping crates. Handling

Fig. 4.1 Machine casting of flat glass.

and loading equipment have rubber suction pads. Cutting operations can also be computer operated without manual intervention.

Gas is usually used to heat the lehr, but sometimes fuel oil is used. The furnace is usually built of brickwork encased in sheet metal. To accelerate cooling, the glass lies on an open track for the last section of its passage.

Use of rolled glass

Rolled glass is used where clear visibility is either not important or not desired. It is translucent with a light transmission of 50–80%, depending upon the thickness and surface texture. It is most frequently used in such areas as bathrooms, washrooms, house and office doors; and for glazing in industrial buildings, railroad stations and elevator shafts. Rolled glass is also popular in canopies, partition walls, partition panels between counters in banks and government offices, and stairway spandrel cladding. It is also suitable for table-tops and for indoor and outdoor light fixtures.

Types of rolled glass

Rolled glass is categorized, according to its appearance, as raw glass, ornamental glass, greenhouse glass, wire-reinforced ornamental glass, and profile glass. There is also colored rolled glass.

Raw glass is rolled glass with the surface rolled smooth or slightly textured. It can be used in applications which do not require transparency or specific optical properties. Raw glass can have a 'hammered' (like metal), coarse-ribbed or faceted surface. Such configurations appear on one surface only. The other is fire-polished, that is, cooled down immediately after being rolled flat.

Ornamental glass is rolled glass with the surface textured by patterned or embossing rollers. A large variety of designs can be given to one or both sides. The designs are not only decorative. Surfaces are often specifically designed for high dispersion and reduction of glare. Glass manufacturers generally have their design names legally protected. Often the name indicates the type of texture used: wave glass, light-scattering glass, rounded (or bullion or bull's-eye) glass, lined glass or trade names such as Karolith or Difulit. Some more nature-related textures such as flowers or frost patterns can also be created.

Greenhouse glass is a translucent rolled glass with a special surface relief to scatter light. It is used in horticulture to scatter incident sunlight and heat evenly in hothouses and garden frames. It should not be confused with horticultural clear glass which is another type of sheet glass manufactured by a different process.

Wire-reinforced glass is made, as mentioned, by embedding wire netting in rolled glass. The embedded wire material can be made of 6 mm mesh webbing, 12.5 mm mesh spot welded netting, or 25 mm mesh honeycomb netting. Common wire-reinforced glass is a raw glass and has limited transparency. It provides a certain protection against break-in and the spreading of fire. For this reason, it is generally found in entrance doors, hall doors, or storage area doors, where its use is officially approved in fire-rated construction elements. When it breaks, the wire mesh holds the glass together to provide safety. Smashed wire-reinforced glass will not fall apart in pieces, thus reducing the risk of bodily injury in accidents involving glass doors and glassed-in elevator shafts. It is used in roof glazing, particularly in factories, where it can carry loads (fallen objects, icicles, snow, etc.) even after the glass is broken.

Wire-reinforced ornamental glass is similar in manufacture and application to plain wire-reinforced glass. The difference lies in the texture of one or both surfaces. The surface design is similar to that found in un-reinforced ornamental glass. Wire-reinforced ornamental glass has practical importance when both decorative effects and safety factors are of concern. It is usually used indoors.

Profile glass is a U-shaped rolled glass for architectural use. Profile glass is manufactured as a continuous glass strip with

Fig. 4.2 Church window made of *dalle* glass.

both edges already bent upwards. The base width is 23–26 cm, and the bent-up height is 4–6 cm. It is mounted in metal frames with a corresponding U-shaped cross section. The joints are permanently and elastically sealed. It can be single or double glazed, and is used for skylights and in wall constructions in factories, warehouses, staircases, or multi-storey garages.

Colored cast glass includes many kinds of cast and rolled glass. There are more than 100 colors of *dalle* glasse (*dalle* is French for 'tile') to choose from. *Dalle* glass is produced in pot furnaces and hand cast into molds resulting in plates approximately 25 mm thick. These are used in special studios to make concrete-glass windows. The glass is broken to the desired size and arranged on a drawing according to an artistic design. The 'composition' is then placed in a steel frame, and concrete is poured into the open spaces and allowed to firm up. These type of windows have become popular in churches (Fig. 4.2). *Dalle* glass is also used to make door handles and all-glass doors (Fig. 4.3).

Fig. 4.3 Door handles made of *dalle* glass.

Opaque glass (opaque, colored flat glass) transmits no light whatsoever. The front side is fire polished and the reverse side ribbed. Opaque glass can be cut, ground, drilled, and otherwise processed (etched, frosted, etc.). It usually comes in white, black, green, gray and beige; there are also multicolored, veined opaque glasses. Since opaque glass is also air- and water-tight it is used as wall-cladding for different exterior and interior architectural applications as well as for furniture and other equipment (shelves and counter-tops). It can be used as wall tiles and mounted with mortar or special adhesives. Large sizes are used as a weather-resistant facade cladding in buildings.

Opaque tiles or panels are resistant to weathering; hairline cracks like those appearing on ceramic clad surfaces do not occur. Opaque glass is very suitable where hygienic requirements prevail, such as in operating rooms, test laboratories and pharmaceutical manufacturing plants. It is relatively resistant to acids and alkalis and unaffected by climatic temperature changes and frost. Limits to its application are imposed by the cost of the special mounting techniques.

Opaline glass is closely related to opaque glass. It is an opaque cast glass with ground and polished surfaces. It has many of the same uses as opaque glass. Opaline glass is preferred when high-quality surface finish is required. It is also used for panels and signs.

4.1.2 Window and plate glass

Only with some difficulties and after years of experimentation were glass process engineers successful in drawing flat glass directly from the tank. Up until 1900 window glass was exclusively made by glassblowers who blew the glass to large cylinders which were then cut lengthwise, put into a special furnace called the flattening furnace, and processed into glass sheets by heating, stretching, 'ironing' and finally annealing into panes of window glass. The first unsuccessful efforts to make window sheet glass by drawing or lifting molten glass directly from the furnace go back to the 1850s.

The Fourcault process

In 1905, a Belgian by the name of Fourcault succeeded for the first time in drawing sheet glass directly from the tank. His process was finally introduced in commercial production in 1914. The most important of the process is the *debiteuse*, a 3-meter long clay block with a slot which floats on the molten glass. When pushed slightly into the glass, the molten glass rises through the slot and is grasped by an iron 'bait'. The thickness of the sheet is irregular initially, but evens out after a while. Rollers draw the solidified glass about 7 meters up the annealing shaft. When it emerges from the shaft, it is cool, annealed and ready to be cut (Fig. 4.4). The slower the drawing speed, the thicker the glass. Up to seven *debiteuses* with annealing shafts can be operated in one Fourcault tank. Major disadvantages of the process are: the fine roller marks on the surface of the glass, and the tendency to devitrify due to the effect of the refractory material of which the *debiteuse* is made.

The Libbey-Owens process

Some time after Fourcault's development, the American, Colburn, successfully devised another method of making window

58   *Flat glass*

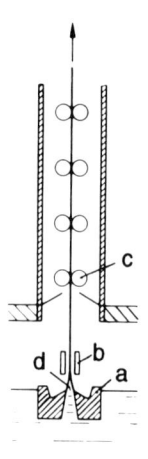

Fig. 4.4 Sheet glass drawing with the Fourcault process: (a) *debiteuse*; (b) coolers; (c) drawing rollers; (d) dome.

glass. With the support of the USA firm Libbey-Owens, the process was further developed. It was given the firm's name when it went into commercial production in 1917. Unlike Fourcault's system, Colburn's process does not work with a *debiteuse*. Initially the glass is drawn vertically from the tank by a metal 'bait'. The edges of the sheet are immediately taken over by cooled side-rollers to prevent the glass from contracting. After travelling about 70 cm, the still-soft glass is guided over a polished steel roller, and turned into a horizontal direction before entering a 60 meter annealing lehr. The drawing speed is twice that of the Fourcault process. Libbey-Owens machines have two work channels from which glass flows in a continuous band, as in the Fourcault process (Fig. 4.5).

The Pittsburgh process

The production system developed by the USA Pittsburgh Plate Glass Company combines the best features of both the Fourcault and the Libbey-Owens processes. It has been in use since 1928. In this process, the glass is first drawn vertically as in the Fourcault process. A *debiteuse* is not necessary, however. Instead, a guidance device made of refractory material is placed

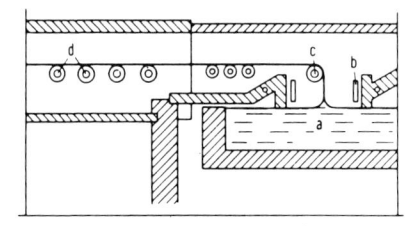

Fig. 4.5 Sheet glass drawing with the Libbey-Owens process: (a) glass melt; (b) cooler boxes; (c) deflection roller; (d) conveyor rollers in the drawing and annealing zone.

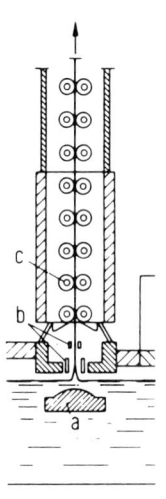

Fig. 4.6 Sheet glass drawing with the Pittsburgh process: (a) drawing guide; (b) cooling zone; (c) drawing rollers.

into the glassmelt at the drawing location. Cooled grippers, shaped like hollow plates, receive the glass. Their slit-shaped cutouts prevent the glass from contracting. After passing through an annealing shaft about 12 meters long, the finished product is cut (Fig. 4.6). The advantages of the Pittsburgh process lie in its production speed, the ability to change glass thickness quickly, and the good glass quality.

Application of sheet glass

Window glass, sheet glass or draw glass (all names for the same product) always has both surfaces smooth or fire polished so no further processing is required after the drawing process. It was the most important type of flat glass for the construction industry for years, but since the 1960s, it has increasingly been replaced by float glass. Window glass is divided into different thickness categories: thin glass (0.9–1.6 mm), standard thickness (1.8 mm), medium thickness (2.8 mm), double thickness (3.8 mm), and thick glass (4.5–6.5 mm). Thin glass is used for picture glazing, for microscope slides, or for protective glass for instruments. Window glass in the next three thicknesses is primarily used in construction. Thick glass was formerly used for such items as store windows and table-tops in furniture manufacturing. Poor quality window glass is used in greenhouses and cold frames (horticultural clear glass).

4.1.3 Plate glass

For a long time, ground and polished plate glass was actually the highest quality flat glass. It was especially suitable for mirrors, where exceptionally distortion-free glass was required. Until the 1960s, plate glass was the predominant base material for glass processing operations in Germany and other countries with highly developed glass industries. The introduction of the float glass process nearly put an end to the old method of manufacturing plate glass, with the exception of 'white' plate glass which is still manufactured in some special glass factories (Other types of flat glass, Section 4.3).

To manufacture plate glass, rolled or thick glass was rough- and fine-ground using grinding compounds on large tables or conveyors equipped with large rotating disks. It was then polished smooth with polishing materials (iron oxide or cerium oxide). Later, large equipment was built in which grinding and polishing was done simultaneously on both sides in a continuous process (the twin process).

4.1.4 Float glass

Since the beginning of the 1960s, the production of sheet and plate glass has been gradually replaced by a modern process, which produces glass of practically plate glass quality, but like sheet glass manufacture this is achieved in a single operation

The production and use of common types of flat glass 61

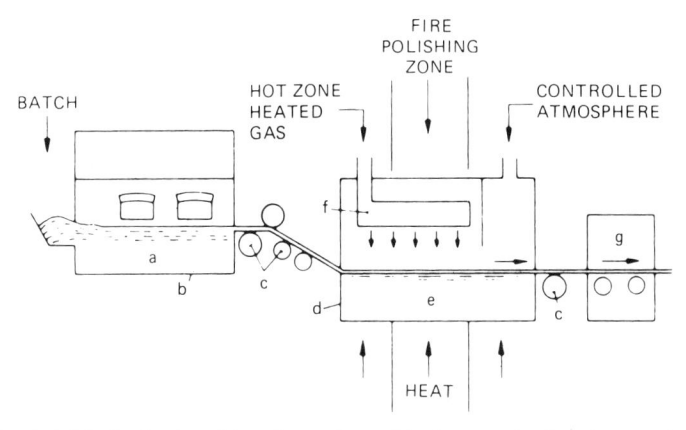

Fig. 4.7 Method of making float glass: (a) glass melt; (b) glass melting tank; (c) transport rollers; (d) float bath; (e) molten tin; (f) heated zone; (g) annealing lehr.

requiring no further processing. This was accomplished after years of research by the British firm Pilkington Brothers Ltd. in St. Helens, Lancashire.

The float process

The verb 'to float' means 'to be buoyant'. And this is basically the principle on which the process is based. Glass type and melting method are similar to those used for other flat glasses. The difference is that the molten glass is fed onto a float bath of molten tin. This tin bath is 4–8 meters wide and up to 60 meters long. To prevent the tin surface from oxidizing with the atmospheric oxygen, the tin bath is held under a slightly reducing protective gas atmosphere. This atmosphere must be carefully controlled since its composition is instrumental for the properties of the contact surface between the glass and the tin which, in turn, influence the thickness of the glass sheet (Fig. 4.7).

The glass floats like an endless ribbon on the tin. At the entrance where the glass first makes contact with the tin surface, the temperature of the liquid metal is about 1000°C. At the exit, it is about 600°C. Tin is the only metal that is liquid at 600°C and does not develop a disturbing vaporization pressure at 1000°C.

Immediately after the exit from the float chamber, special rollers take up the glass and feed it into the annealing lehr from which it exits at about 200°C. After cooling to room temperature on an open roller track, it is cut, packed, and stored either for later shipment or for further processing into such products as safety glass or double- (or multi-) glazed units.

Float glass can be made in thicknesses between 1.5 and 20 mm. There are two techniques to accomplish this. To produce thin float glass, rollers control the width and speed of the glass ribbon. For thick float glass, the glass floats against graphite barriers, so that the ribbon flows out thicker. Thus the desired widths and thicknesses are achieved. The output of a fully automated plant can reach 3000 square meters per hour requiring an operating staff of only four. Tinted, but still transparent float glass (known as chameleon glass) can be produced by doping the glass with the metal indium and then subjecting the hot glass to electrical impulses. This process, known as pulse electro-float, was developed by Pilkington Brothers Ltd. It is a modification of the electro-float process used to produce tinted solar control glass, in which particles of an alloy of copper and lead are dispersed electrolytically on the surface of the glass while it is in the viscous state and then reduced to colored metal particles in a hydrogen/nitrogen atmosphere.

Today, there are numerous float glass plants in operation worldwide (Fig. 4.8). The production of plate glass has been all but abandoned in Europe and has been replaced by the float process. Sheet glass is still drawn in several places, primarily using the Fourcault process which is the oldest fully automated method of producing drawn flat glass.

Applications of float glass

Since it combines the qualities of plate glass and sheet glass, float glass is used for glazing wherever clear transparency is required in buildings. It is also a base material for safety glass, as required primarily in the automotive industry and for other means of transport. Mirrors, and otherwise finished or processed flat glass for furniture or interior decorating, are also made from float glass. In addition, there are numerous other uses in

Fig. 4.8 Float glass plant of Flachglas AG in Weiherhammer, Germany.

precision mechanics, especially where extreme surface flatness is required, as for example, in visual displays.

4.2 TECHNICAL IDENTIFICATION OF SODA-LIME FLAT GLASSES

Table 4.1 contains important physical and technical data pertaining to normal flat glass. Similar values hold true for common soda-lime container glass.

4.3 OTHER TYPES OF FLAT GLASS

In addition to rolled, sheet, plate and float glass, there are a number of other flat glasses. Most of them have very specialized applications and are subject to stringent technical requirements. For this reason, they are generally considered to be special glasses, a subject to which a separate chapter is exclusively devoted. But there are several other types of flat glass in use in daily life which are worth mentioning here. Most of these glasses are made by glass companies which belong to the Schott Group.

As already mentioned, near-colorless, temperature-resistant glass is obtained by using raw materials which contain the least possible amount of iron. Borosilicate flat glass is especially important for the consumer markets under the designations *Tempax* (Schott) and *Pyrex* (Corning, USA). Tempax is manufactured by a special drawing process, while Pyrex is rolled. Panels of up to 2 square meters manufactured by these

Table 4.1 Data for normal flat glass

Density	2.5 g/cm^3
Thermal expansion (linear)	8.5–9.5 × 10^{-6}/K
Thermal conductivity (k-value)	
at 5 mm thickness	5.6 W/m^2 K
Modulus of elasticity	~7 × 10^4 N/mm^2
Bending strength	~30 N/mm^2
Specific electrical resistance	10^{12}–10^{13} Ω cm
Transmission of light between 0.36 and 2.5 μm wavelength (visible spectrum and near infrared) at 4 mm thickness	~75–90%
Viscosity data	see page 21
Transformation temperature (~ 10^{13} dPa s)	525–545°C
Softening point (10$^{7.6}$ dPa s)	710–735°C
Working point (10^4 dPa s)	1015–1045°C

processes usually have fire polished surfaces. If smoother surfaces are required, the glass has to be ground and polished conventionally.

Owing to the high thermal resistance and temperature shock resistance, these glasses (Other hollowware, Section 5.7; Borosilicate glasses, Section 6.2) are used as door windows in household stoves, ovens, washing machines, and the like. They are also used for colorless substrate glasses in hot panels with electrically-conducting surface layers, and for heat-absorbing filters and mirrors.

The flat glass made by Deutsche Spezialglas AG in Gruenenplan, Germany merits mention as a completely colorless plate glass. This glass, identical in composition to ophthalmic glass, is processed like classical plate glass, i.e. it is ground and polished. Drawn sheets of 0.05–0.2 mm thickness are also manufactured. Thin microscope cover slides are an example of the use of this alkali-boro-zinc-silicate. The temperature shock resistance of these glasses is equivalent to that of common flat glasses.

4.3.1 Antique glass

Flat glass of the highest purity and clarity is not always desired. Antique glass has the deliberate appearance of pre-

Fig. 4.9 Antique glass sheets.

industrial age flat glass (Fig. 4.9). The surface texture is uneven and there are bubbles inside caused by incomplete refining or by the addition of gas-producing substances to the glassmelt shortly before its processing. Depending on the production method, antique glass can be classified either as a sheet or as a plate glass. It can be produced either by the original (or antique) method from mouth-blown cylinders or drawn by machine. If only one side of the glass has a surface texture, it is called neo-antique glass. These designations have been generally adopted, even though they are misleading. In a strict sense, none of these glasses is antique, or actually old. They are rather glasses manufactured with antique methods. Mechanically drawn antique glass shows a repetitive regular pattern.

Antique glass is popular for rustic or nostalgic style decoration, in which case it is often glazed with lead or brass strips into furniture and windows. In living rooms or restaurants decorated in the old German style, antique glass is regularly used.

So-called rolled antique glass is different from neo- or genuine-antique glass. It is an ornamented rolled glass with antique-glass-like surface texture.

4.3.2 Flashed glass

Not all colored glasses are tinted in a batch process. When a clear glass is covered by a layer of colored or opal glass, it is called flashed glass. The flashing technique is used for flat glass and hollow glass (lamp globes, tableware and decorative glass). Otto Schott introduced it for technical use.

Flashed opal is the best known glass in this category where the traditional mouth-blowing technique is replaced by a mechanical process. In this process, an endless sheet of clear glass is drawn from a Fourcault tank through a *debiteuse*, like sheet glass. At the same time, a thin sheet of opal glass with similar thermal properties is drawn from a small tank located nearby. The fusing of both sheets takes place where the clear glass emerges from the *debiteuse*. Glass thicknesses between 1.5 and 7 mm can be achieved using this method.

Flashed opal glass is well-suited for evenly dispersed, glare-free room illumination, for translucent room dividers, for bathroom windows, for shelves in showcases, X-ray picture viewers, or as dial panels for technical and medical measuring instruments (Fig. 4.10).

4.4 PROCESSED FLAT GLASS

In addition to good optical quality (undistorted transparency), strength and weather resistance, requirements are often placed on flat glass which cannot be met satisfactorily, if at all, by changing the glass composition. Architectural glazing is a case in point, where it is necessary to have certain qualities such as heat insulation, light and sound protection, and increased resistance to breakage. Special safety is required when flat glass is used in vehicles. Flat glass can also be given considerably altered optical and decorative properties by special surface treatment. All these property changes are achieved by industrial processing of flat glass.

4.4.1 Glasses with altered radiation, heat and sound transmission characteristics (solar, thermal and sound insulation)

Simple construction glass transmits 75 to 90% of incident light depending on the thickness of the glass (see Table 4.1).

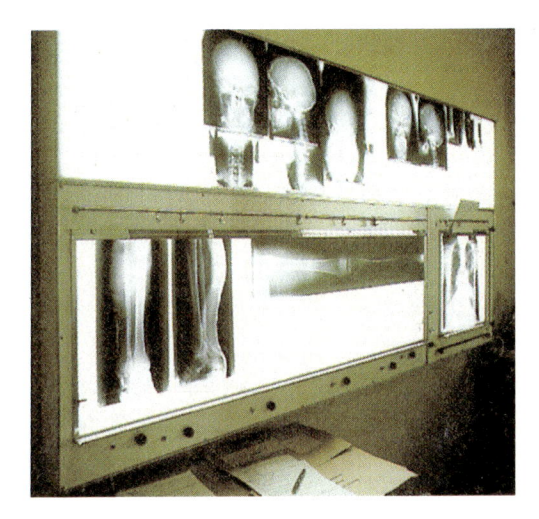

Fig. 4.10 Flashed opal glass in X-ray film viewers.

Transmission decreases substantially at incident angles above 40° because of the increasing degree of reflection. The amount of heat transferred through glass depends on its thermal conductivity which is, in turn, dependent on the temperature difference between the inside and outside surface of the glass, and on the thermal radiation into the cooler side surrounding the glass. These two properties are identified by the thermal transfer coefficient (k). It gives the amount of heat per square meter per second which passes through the glass at a temperature difference of 1°C. It is measured in $W/m^2 K$ (= 0.86 kcal/ $m^2 h$ °C). Together with the size and position of the window, it determines the energy consumption for heating and air conditioning of buildings. It also has a considerable influence on the climatic comfort of interior spaces. Advantageous changes in the radiation transmission and k-value of flat glass can be achieved in different ways. There are three basic product categories in use:

1. multipane glazing;
2. glasses with increased absorption or dispersion of visible and infrared light; and
3. flat glass with altered surface properties.

In many cases, a combination of these products is used (Fig. 4.11).

In Group 1, insulating glasses are the most important. They consist of two (sometimes three) panes which are either fused directly to one another at the edges or are glued or soldered to a metal frame in such a manner that the space between the panes (9–12 mm) is airtight. A desiccant inside prevents the inner glass surface from fogging as a result of temperature change. The thermal conduction of the enclosed air – which is sometimes replaced by gases having a low thermal conductivity – is substantially reduced due to the narrow space. The k-value of a double-glazed insulating window having 6 mm thick panes and 12 mm air gap is only 3.1 W/m^2 K as opposed to 5.6 W/m^2 K for single pane glazing. More advanced developments in double glazing windows achieve 1.2 W/m^2 K. This improved insulating effect is a result of coating the outside surface of the inner pane with a thin layer of Ag or Sn. Further improvement of the k-value can be obtained by filling the gap between the panes with an inert gas, e.g., argon. Under extremely good conditions, k-values of down to 0.7 W/m^2 K are reached and with solar heat incident, even a heat gain can be achieved.

Insulating glass windows also offer good sound insulation and help reduce street noise. The sound attenuating effect of a window is measured in decibels. This unit expresses by how many 'phons' the sound level is reduced when travelling from the outside through a window to the inside of a room. A single pane window has a sound attenuating effect of about 25 decibels, while that of an insulating window achieves 30–40 decibels. Loud street noise is dampened by almost a half. The effectiveness of insulating glass is also dependent on other factors, such as the angle of incidence and the sound frequency.

Among the flat glasses with increased absorption listed in Group 2, glasses with additional bivalent iron oxide content have become important because of their low transmittance in the near infrared. From incident sunlight the heat radiation in the 0.7–2.5 μm range is absorbed. As absorption sharply increases above 0.6 μm, the transmitted white light shows a noticeable green tint, which can be corrected to a certain extent by the addition of other absorbing components. The k-value of

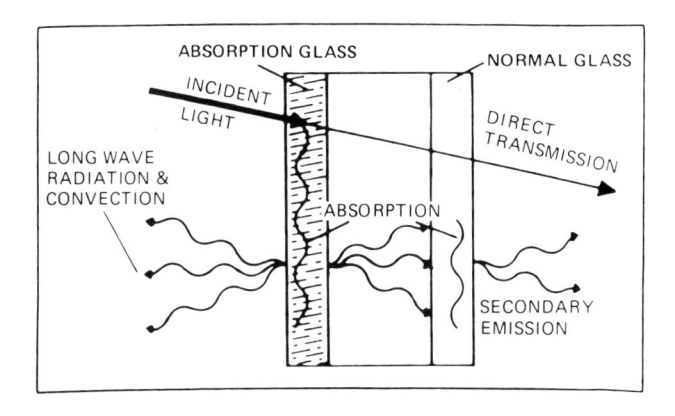

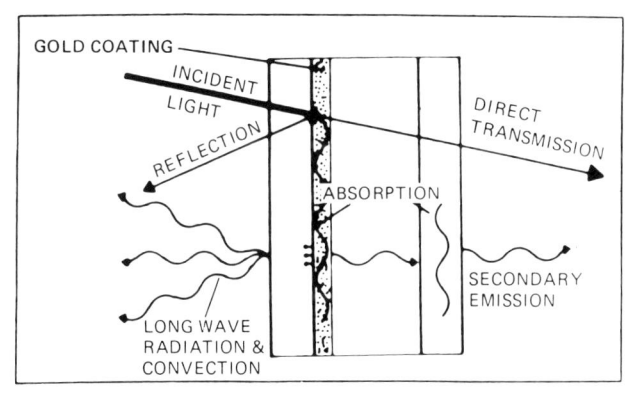

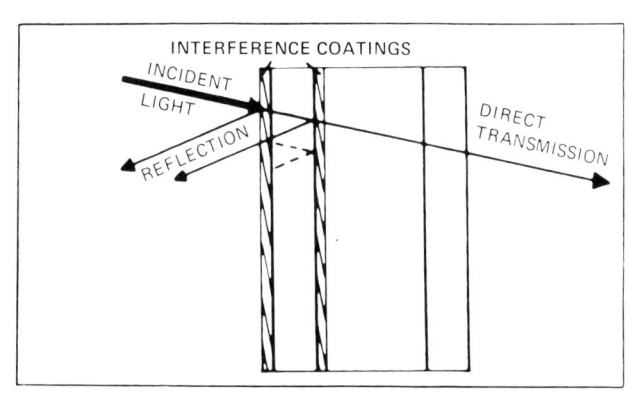

Fig. 4.11 Radiation distribution in a double pane insulating window with absorbing and reflecting panes.

the glass, on the other hand, is not noticeably influenced by the absorption properties. In sunlight, the glass warms up to a temperature at which the heat radiation into the environment is equal to the energy gain of the absorbed radiation. Such glass offers limited sun protection as building glazing. It is much more effective in vehicles due to the cooling effect of the wind. Absorbing glass must be tempered when used in buildings (Safety glass, Section 4.4.5) because of the danger of breakage caused by uneven heating of the window surface (if it stands partially in the shade, for instance). Similar problems arise with other tinted glasses used in building glazings.

To reduce the brightness of large windows, gray and bronze color tinted glasses were developed. There are also special yellow tinted protective glasses on the market which are opaque to ultraviolet and shortwave blue light in order to inhibit the fading effect of sunlight on fabrics, paintings, etc.

Double panes can be laminated with a plastic film containing absorbing materials in order to reduce transmission. This is how windshields with a gradually increasing tint towards the top are made to reduce glare. Ultraviolet radiation can also be better eliminated with an embedded ultraviolet-absorbing film rather than with tinted glass. The primary importance of these laminated glass panes lies in their use as safety glass (Safety glass, Section 4.4.5). Light diffusing, translucent double-paned glasses are also manufactured with fiberglass filled spaces and with edges sealed like those found in insulating windows. Such a product is especially useful for thermal insulation.

Group 3 offers numerous possibilities to alter the radiation and thermal transmission of glass. Using thin-film technology to coat large surfaces, the reflection and absorption characteristics of glass surfaces can be altered in such a manner that requirements in energy saving and optimum air conditioning can be widely achieved. For solar control, regular flat glass can be coated with metallic and/or oxide layers so that with sufficient light transmission, the heat effective radiation is largely reflected (40–60%). The effect is partially based on the high reflectivity of precious metals and partially on optical interference effects of highly refractive coatings, the thickness of which are usually about a quarter of the mean wavelength of the radiation. The coatings are normally applied by vapor

Fig. 4.12 Hessische Landesbank in Frankfurt/Main with IROX solar reflective glass.

deposition, by sputtering under vacuum, by dipping into or by spraying with solutions, followed by firing.

In order to improve their insufficient mechanical resistance, vacuum-deposited precious metals require a highly abrasion-resistant protective coating. Among the materials used to accomplish this are special borosilicate- and calcium-aluminosilicate glasses (from Schott) which provide highly durable protective coatings when vaporized by electronic bombardment under vacuum. Electroless metallization processes (Reflective flat glasses, Section 4.4.3) are also used, particularly for the production of solar protection coatings. The usual configuration for a double-glazed solar protection insulating glass combines a coated pane with an uncoated pane. This way the energy consumption in air-conditioned buildings can be substantially reduced (Fig. 4.12).

In order to improve the insulating performance of glass at low temperatures, i.e. to reduce the k-value of windows, thin-film coatings can be used provided they have a high electron conductivity. Precious metals and several oxidic semi-conductors (tin oxide, indium oxide) show this property. They are deposited either by vacuum processes or by precipitation of vapor-state or liquid compositions combined with a simultaneous or subsequent heating process. The coatings allow more than 70% light transmission but increase the reflectivity in the infrared range to up to 90%, so that the thermal emissivity of the glass falls to between one-fifth and one-tenth of the normal value, thereby significantly reducing the k-value.

Instead of coating the glass itself, cost considerations have brought about the use of appropriately coated plastic films which adhere to the internal surface of the glass. Up to now, metallized films (with Al, Ag or Au) have been applied almost exclusively. Their light transmission usually lies below 50%. Vacuum coating processes are mainly used to metallize the films as they are economical and require a minimal amount of heat.

4.4.2 Non-reflective glasses

The view of objects behind glass, such as pictures, display items, measuring instruments, etc., is often disturbed by the light reflections at the glass surface. About 8% of incident light under a 90° angle is reflected in such cases, and it increases with the angle of incidence. This can be eliminated in several ways, such as by fine etching (silk matting) the surface, which diffuses the reflected light, or by applying non-reflective interference coatings with defined refractive indices and thicknesses. Interference coatings are applied by usual coating methods or by leaching techniques. Silk-matt etched glasses are suitable for pictures only if they are contact-mounted. Otherwise, the contrast is impaired by the scattering effect, which increases with the distance between the glass and the picture. It is widely used as anti-Newton-ring glass in slide projectors, whereby the matted surface inhibits the appearance of Newtonian interference colors upon contact with the slide.

Leaching can only be carried out successfully on certain special glasses and is therefore rarely used. Two examples for its use are large anti-reflection blocks of radiation shielding glass and optical glasses exposed to very high intensity laser beams (Special optical glasses, Section 6.6.4). On the other hand, regular flat glass up to several square meters can be coated on both sides with a triple interference coating process developed by Schott. This reduces the visual reflection of light to about one-tenth of the normal value. These layers are produced in a sol-gel process (Section 3.3.8). Since the coating causes no noticeable absorption or dispersion loss, the reduction of reflection is combined with an increase in light transmission of the equivalent amount. This effect is important in optical systems such as camera lenses. Such anti-reflection coatings are almost always vacuum-deposited. They are also used to reduce reflection and glare in spectacles.

4.4.3 Reflective flat glasses

A high-quality defect-free flat glass is required to make mirrors that offer the observer a natural, undistorted image. Float glass is the most economical glass used to meet these requirements.

The manufacture of mirror glass

The flat glass panel is first washed and then the surface that is to be coated is treated with a stannous chloride solution. This facilitates the silvering process by activating the glass surface. A solution made of silver nitrate, ammonia, caustic soda or caustic potash, and distilled water together with a reducing agent made of dissolved glucose is then sprayed onto the surface of the glass. The reducing agent causes the formation of silver nuclei which combine with the tin from the pretreatment forming a solid, fine crystalline silver film. This coating is opaque at thicknesses of more than $0.5\,\mu\mathrm{m}$. The silver layer is usually allowed to grow to a minimum thickness of 0.01 mm. In a similar process, the silver is then coated with a layer of copper. After drying, two coats of protective lacquer are applied. Today, this is all accomplished in fully automated processes. In the past, the silvering solution was deposited by repeatedly pouring it over the glass.

The electroless metallization of glass by wet processing is similarly applicable for gold and copper coatings. In strongly reducing solutions (preferably containing hypophosphites) it is even possible to deposit metals of the iron group (particularly nickel and nickel–copper alloys) as mirror surfaces. The attainable degree of reflection is generally less than 50% for such coatings. Similar processes are used in the USA to manufacture solar protection glass. Very thin coatings with a 15–30% light transmission and a 40–25% reflection are used for this. The remainder of the radiation in such glass is absorbed by the coating.

The use of mirrors

Most mirror products are used in the home. Bathrooms and toilet areas are usually equipped with mirrors. Bedroom furniture usually includes mirrors. Decorative mirrors can also be made of rolled glass, both colored and tinted. The coating on antique mirrors is intentionally cracked so that it looks old. Mirrored walls have an enlarging effect in small rooms. Rear view mirrors in automobiles contribute to our safety in traffic every day. By applying several oxide coatings in a dip coating process, semi-mirrors based on optical interference are made for anti-glare rear view mirrors and one-way mirrors, allowing observation of brightly lit rooms from darkened areas. Self-service stores often have these devices as a monitoring system (Fig. 4.13).

At present, the best type of outside rear view mirror for cars is the aspherically curved low-glare mirror developed by DESAG (Deutsche Spezialglas AG). The rear view angle is 48° (as opposed to only 16° for regular flat left side mirrors) and glare is reduced to 35–45% without substantially diminishing the brightness of the objects in the mirror. The mirror has a scarcely noticeable constant curvature over about two-thirds of its surface. Only the remaining third shows a curvature with a continuously decreasing radius, increasing the rear view angle, thus eliminating the 'blind spot'. The low glare is obtained by depositing several metal oxide coatings on the front surface and 'burning' them in at approximately 400°C.

In the field of solar energy utilization, mirrors of various forms and designs are often used as solar energy collectors.

Cold mirrors are manufactured by applying at least 12 optical interference coatings with alternating high and low refractive indices onto a heat-resisting glass. They are calculated in such a fashion that only visible light is reflected to a great extent,

Fig. 4.13 Heated anti-glare rear view mirror.

while thermal radiation passes through or is absorbed. Chemical reduction processes are suitable only for back-side mirror coatings as the layers are insufficiently durable without a good protective coating. Optical mirrors are either front-surface-coated with stable vapor deposits to prevent double reflection; or rear-surface-coated with the front surfaces anti-reflection coated.

4.4.4 Other surface finishing techniques for flat glass

Many surface treatments for flat glass are in response to fashion trends. For a long time, frosted designs were popular. To achieve this effect, hydrofluoric acid is poured onto the surface and then quickly rinsed off. Since the acid does not attack the glass evenly, designs resembling ice-ferns appear. A similar design results when glue is applied to a matt glass. The drying, shrinking glue chips off and tears off thin layers of glass, leaving a frosted design.

In etching with hydrofluoric acid, different degrees of frosting or matting are achieved. Frosted glass is used in interior designs and in windows where transparency is not desired.

Sandblasting can be used to make varying degrees of rough or matt glass surfaces. By partially covering the surface beforehand with stencil masks, the glass surface can reproduce representational or abstract decorations.

A further development is the engraving of flat glass using a grinding wheel at the tip of a flexible driving shaft. This permits the engraving of free-hand designs onto the glass surface.

4.4.5 Safety glass

Everyday experience shows that flat glass often breaks under a slight mechanical load such as impact or pressure as well as under quickly changing temperatures. The sharp fragments of broken glass can cause serious injury. Special treatment can both render the glass less susceptible to breakage and considerably reduce the danger of injury. The result of such processing is safety glass.

Tempered (or toughened) glass

In this process, pre-cut pieces of flat glass are vertically suspended above or placed horizontally into processing equipment where they are quickly heated to about 150°C above the transformation temperature (Technical identification, Section 4.2). Immediately upon exiting from the furnace, the glass is chilled with cold air from an appropriately designed system of air jets. As a result of this fast cooling, the glass surface is 'frozen' in an expanded grid structure whereas the glass inside cools off more slowly, allowing much more shrinkage in the structure. Since they are bound together, the outer surface layer is subjected to compression and the inside to tension. The procedure itself is called thermal or physical toughening of glass; the finished product is called tempered glass or single pane safety glass. The amount of pre-stress in the glass surface depends on the thermal expansion, the modulus of elasticity and the transformation temperature (T_g) of the glass as well as on its thickness and the amount of heat transmission between the cooling medium and the glass surface. In most cases a pre-stress of 80–120 MPa (1 mega-pascal = 10^6 N/m^2) is desired. This increases the bending strength to two to three times the value of that of an untempered glass. When thermally toughened safety glass is damaged, e.g., when a windshield is hit by a stone while the car is moving, or if a car is involved in an accident, the glass breaks into many small, almost regularly shaped pieces with no long sharp cutting edges. The size of the particles of glass can be controlled in advance by modifying the

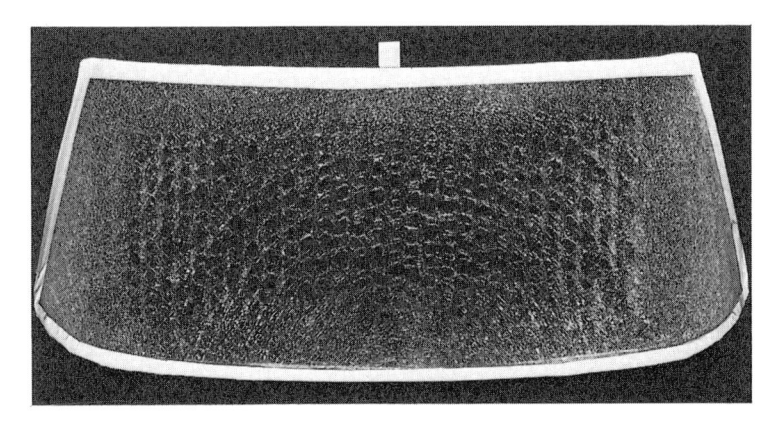

Fig. 4.14 Breaking pattern of a tempered safety glass windshield.

toughening criteria. This is important for automotive windshields in order to maintain the driver's visibility in case of breakage. Often single pane safety glass for windshields has so-called vision islands which crack into larger pieces than the remainder of the glass. They provide reasonably good vision until the damaged windshield can be replaced. There are several DIN-German Industry Standards (German Standards Institute) test procedures for single pane safety glass (Fig. 4.14).

It is soda-lime float glass that is usually toughened for automotive applications. Single pane safety glass is usually produced by the large flat glass companies which manufacture the raw glass itself. During the same process, the tempered glass is also bent, since front, rear, and side windows are often curved. Bending must precede tempering since the temper would disappear at the temperature required for the bending operation. In office buildings, tempered flat glass is used in glass doors, room dividers, elevator glazing or stairway landings. Ornamental rolled glass is also popular in applications where clear vision is not desired (the possibility of processing safety glass into multipane glazing has already been discussed on p. 70).

The curtain walls of facades of large buildings are often covered with tempered flat glass for decorative purposes. The back of these glass panels is often coated with a color-coordinated layer of enamel. To achieve this, the glass is dusted with enamel frit before entering the toughening furnace. The enamel melts when heated and bonds to the glass surface without influencing the prestressing process. This process results not only in the desired decorative effect but also in a greater weather resistance. In the domestic appliance sector, toughened glass is used for control panels incorporating touch-sensitive switches, which simplify the operation of washing machines, dishwashers, electric ranges, television sets and elevators. Their use can provide appliance manufacturers with greater design freedom.

Thermally toughened glasses cannot be further processed except to reduce the width and length by grinding. Therefore, the glass must be cut to final size prior to tempering. The same applies to drilling and edge work. This is especially important for glass doors which have to be equipped with hardware such as hinges, locks and door handles.

Chemical toughening

The chemical prestressing of glass consists of creating a compressive stress in the glass surface by modifying the surface composition without changing the inner body of the glass. In 1891 Otto Schott succeeded in producing a very strong glass by coating a glass of high thermal expansion with a glass of low expansion. This process is related to those used today in the production of tableware and flashed opal glass.

Chemical prestressing processes are based on ion exchange. The ions of the individual chemical elements have different radii and are arranged at different distances from one another. For example, if a sodium-containing glass is slowly heated to just below the transformation temperature in a molten potassium salt, the sodium ions migrate from the glass into the salt melt, and the potassium ions leave the salt melt and enter into the glass surface. This is called ion exchange. Since the potassium ions which penetrate the glass have a 30% larger radius, a shortage of space arises in the glass surface, which results in a compressive surface layer. The exchange zone must be at least 0.1 mm thick in order to increase the strength by five to eight times. Glasses toughened by ion exchange are used for special purposes in the aircraft industry, for centrifuge tubes, and in the lighting industry. The same process is used to toughen ophthalmic lenses.

Laminated glass

Laminated (or compound) safety glass consists of two or more panes (usually float glass) which are joined with a viscous plastic layer. The solid joining of the glasses occurs in a pressurized vessel called an autoclave where under simultaneous heating of the pre-processed 'sandwich' the lamination takes place.

When laminated safety glass breaks, the broken pieces of glass stick to the internal tear-resistant plastic layer. The pieces do not break away and the broken sheet remains transparent.

Unlike prestressed safety glass, laminated glass can be further processed. It can be cut, drilled and edge-worked.

Laminated safety glass is used in building windows, where break-in or escape hampering or explosion protection glass is required. It is also used in spandrels, in walls, room dividers and roofs. There is a clear distinction between safety glasses that are impact resistant, breakage resistant, bullet proof and explosion effect dampening. In many countries law requires that auto windshields be made of laminated glass. Even where legislation is not in place, it is still widely used. It is used as a protective glass in machinery, instruments, and television sets. In addition, there are several special forms of laminated glass which should be mentioned.

Armor plate glass consists of at least four laminated glass panes and is at least 25 mm thick. Such glasses are resistant to handgun fire.

Armor plate glass over 60 mm thick is bullet proof. Cashier's offices, bank counters, jewelry store windows, or armored transport vehicles can be equipped with armor plate glass (Fig. 4.15).

Wire compound glass describes glass with steel filaments embedded between two sheets of glass to increase safety.

Alarm glass is, to an extent, a further development of wired compound glass. Embedded in the intermediate layer are

80   *Flat glass*

Fig. 4.15 Handgun-fire test on armor plate glass.

0.1 mm thick wires which form an electrical circuit. When this glass breaks, the circuit is interrupted setting off an alarm at the same time.

Transparent heated glass is useful against fogging and ice formation. This can be the case on windshields and rear windows in cars, in airport control towers, in flower shops, in cold storage vaults, or in indoor swimming pools. This effect is achieved in two ways. Either the intermediate layer contains barely visible wires which heat up when connected to an electrical power source, or the glass is coated with a transparent electrically conducting surface layer which heats up similarly.

Opaque heated glass is clear or tinted float glass which can have an electrically conducting silver circuit printed on the reverse side with, if required, a decorative pattern on the visible side. During the toughening process (see Tempered or toughened glass, above) the printed circuit and pattern are fired into the glass. The electrical circuit can then be coated with a protective ceramic or lacquer paint. Special soldered pads facilitate the wiring of electrical connections. Output can be up to 500 W per 0.1 m². However, at over 250 W per 0.1 m², a thermostat is necessary to limit the surface temperature to a maximum of 200°C. Panels up to 1200 mm × 800 mm and 4–8 mm thick are available (Schott 'heatable' *Durax*). Primary applications are in the food warming trays for use in homes, restaurants and canteens, serving trolleys and kitchen countertops, decorated heating panels in bathrooms, etc.

Antenna glass is also produced in two ways. In one method a thin wire of a given length and shape is embedded in the intermediate layer of a laminated windshield. In the other method, the antenna function is carried out by a conductive silver thread which is fused into the inside surface of a windshield.

Colored laminated glass (both fully and partially colored) is used in buildings and vehicles. The films between the sheets of glass are colored. They can have heat reducing, ultraviolet absorbing, reflecting, or anti-glare qualities (Section 4.4.1 above).

Laminated glass used in combination with other glass

The risk of injury becomes smaller when the inside sheet of a laminated glass is thinner (1.5 mm, for example). The same purpose can be accomplished by using a tempered glass on the inside. If safety is to be combined with heat and sound insulation, laminated glass is used in combination with normal glass or another laminated glass to make insulating glass. Very thin laminated glass is also used in safety spectacles and helmet visors because it is less susceptible to scratching than plastic.

As widely differing glass types can be combined in multipane glazing, there are many practical applications where a combination of several types of processed flat glass are used. Insulating safety glass is found, for example, in railroad cars, sometimes in conjunction with sun protection glass. Gatekeepers' houses can be equipped with multipane glazing containing safety glass or even gunfire-resistant glass. Airport control towers are glazed with large areas of glass to provide maximum vision. Air traffic controllers are assured of comfort-

Fig. 4.16 Fire-resisting glass under test.

able climatic conditions through the use of a sun protection glass as one component in multipane glazing.

4.4.6 Fire-resisting glass

The increased use of glass in buildings (in facades, spandrels and partitions) also increases the possible hazards when a fire breaks out. Normal flat glasses shatter after only a short time of exposure to heat from one side. Large pieces of glass fall out of windows, and the danger of fire spreading quickly to another level increases. So far, there has been only one method of retarding the shattering of glass windows during a fire, i.e. a wire inlay inside 6–8 mm thick panes which prevents the glass from falling apart when it breaks.

Like other construction materials, fire-resisting glazings are classified into fire-resistant classes G and F in defined fire tests. According to the test described in the German Industry Norm Standard DIN 4102, Part 5, glazings which resist flames and smoke from spreading for 30, 60, 90, 120, or 180 minutes are categorized as G 30, G 60, G 90, G 120 or G 180. Likewise classes F 30, F 60, F 90, F 120 and F 180 represent glazings which, in addition impede radiant heat from spreading for at least 30, 60, 90, 120 and 180 minutes respectively, and which do not heat up on the side unexposed to fire to more than an average of 140°C above the initial temperature. Additionally, such glazing must withstand certain steel ball impact tests (Fig. 4.16).

According to this classification system, the above-mentioned wired glass with a concrete frame and with less than $2\,m^2$ surface area conforms to class G 60. For optical reasons, applications are mainly limited to partition walls, doors and skylights.

New developments in fire-resisting glass have resulted in wire-free glasses suitable for windows. *Pyran* (from Schott) is a thermally prestressed borosilicate glass which is officially approved for class G 90 in $2\,m \times 1\,m$ sizes and class G 120 in $1\,m \times 1\,m$ sizes. Because it is prestressed, it breaks into relatively small pieces upon impact. Several flat glass manufacturers have also introduced double-glazing systems into the market, the inner space of which is filled with intumescent material which acts as a heat shield in case of fire. They conform to classes F 30 to F 90.

5
Hollowware and glass tubing

The most widespread glass products belong to the hollowware family. Hollowware is encountered everywhere. In the broadest sense, these products are consumer goods such as bottles, or consumer durables such as drinking glasses or glass lamps.

Most hollowware is made of soda-lime glass. Crystal and lead crystal and numerous other special purpose glasses are exceptions.

The processing and finishing of hollowware are important fields in their own right creating a wide variety of products. Their starting material is hollowware manufactured by the glass factory.

5.1 THE MOST IMPORTANT TYPES OF HOLLOWWARE

The huge hollowware sector can be divided into several groups, according to different criteria. One is the manufacturing process, which can be subdivided into mouth blowing, machine blowing and pressing, to name the most important ones.

It can also be classified according to the chemical and physical properties. This results in such classifications as bottle glass, semi-white hollowware, crystal glass, lead crystal and so on.

Usually hollowware is classified according to its use. This is how worldwide statistical data in the industry are compiled. There are container glasses (bottles, jars, medical, and packaging glass), tableware (drinking glasses, and other glassware for household use) and construction hollowware (glass building blocks, etc.). Medico-technical hollowware and illuminating glasses fall primarily within the realm of the special glasses (see Chapter 6) and are discussed there.

Each of these types of hollowware will be discussed individually. However, since many of them use the same manufacturing processes, the following section will first deal with this subject as a whole.

5.2 THE SHAPING OF HOLLOWWARE

The first known hollow glass body was fashioned in ancient Egypt by coating a shaped sand core with viscous glass (Fig. 1.1).

The real beginning of hollowware manufacturing did not start until the invention of the glassmaker's blowpipe in about 200 BC. The technique of shaping glass using a blowpipe has survived in the originally developed form to the present day. Even the operation of modern glassblowing machines is based on the manual techniques, although they now may hardly be recognized.

5.2.1 The mouth-blowing process

The glassworker's blowpipe is a steel tube with a wooden handle and a mouthpiece at one end. It is about 1.5 meters long. At the other end is the nose (or gathering head), an extension which picks up a gather when dipped into molten glass. By rotating and swinging the pipe, the glass is prevented from running off while it cools. The gather can be dipped into the melt again to pick up more glass, depending on how much is required (Fig. 5.1).

A small hollow body, the parison, forms by giving a short puff into the pipe. Its external shape can be formed by turning it in a molded, water-soaked beech wood block (blocking) or rolling it on an iron plate, whereby the gather becomes more viscous as it cools.

The subsequent reheating in the furnace, along with rotating and swinging of the pipe allows the temperature difference to equalize before the product takes its final shape (Fig. 5.2).

The finishing step of individually shaped articles is done completely manually, using only a few special pieces of equipment such as a wooden board (paddle) and tongs. However, the glass is more often blown into a mold which allows the production of many identical hollow items.

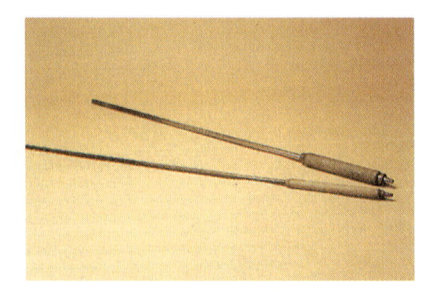

Fig. 5.1 Glassblower's pipes.

Fig. 5.2 Glassmaker at work.

Originally made from water-soaked beech wood, molds are also made of graphite or cast iron for large volume production. A coating mixture of sawdust and a binding agent is 'pasted' on the inside surfaces of cast-iron molds. This paste is burnt in and sprinkled with water before each use. The paste lasts for about 500 operations, after which it must be renewed. The life of a metal mold is practically unlimited.

On contact with the rotating glass, an isolating steam cushion forms in the water-soaked mold; and due to a slow cooling process, a hollow body with evenly thin walls can be blown. This results in a brilliant smooth surface such as the one obtained in free mouth blowing, which is unrivaled by any other forming techniques (Fig. 5.3).

Other forming steps which must be carried out before hollowware is ready for use include shaping and attaching of

The shaping of hollowware

Fig. 5.3 Formfree finishing.

Fig. 5.4 Attaching a handle.

another glass-gather to form a handle, a stem, or a base; the separation from the pipe; the transfer into a lehr; and cutting off the excess glass (Fig. 5.4).

The various steps in the manual production of hollowware are performed by specialized craftsmen, meeting the highest demands of physical condition, manual dexterity, and artistic talent. Thus one can more easily understand that only high quality tableware (crystal goblets, for example) and low quantity technical articles of intricate shapes are still being made by hand today by an ever-decreasing number of glassmakers. The low cost of producing a multitude of hollowware products which we encounter in every day use (beverage bottles and light bulbs, for example) is achievable only by large-scale automated production.

5.2.2 Machine blowing

The development of mechanical production processes led to the first automated blowing machines, made by Michael Owens in 1903. Since then, their further development has been an ongoing challenge.

The first machine-made products were bottles (packaging glass). Owing to the lower surface quality requirements, no paste molds were necessary. This made fast heat dissipation possible, thereby increasing machine output. The hourly output increased from about 17.5 to 90 pieces per finishing mold, and it has risen to about 900 and more in the most modern machines. In later years, the rotating mold (paste-mold) process also was successfully automated.

All automated blowing processes follow the basic manual glassblowing steps after formation of the gather: (1) preforming (forming the parison); (2) reheating (equalizing the temperature differences); (3) final blowing. Although there are many different machines in use today, they can be classified into three basic types:

1. continuous rotary machines, used partly to manufacture container glass, but primarily in paste-mold processes;
2. side-by-side, complete sections, each independently driven, like the I-S machine (Independent Section machine), which dominates the container glass manufacturing industry (Fig. 5.5);
3. the chain or ribbon-type machines, which are used to make light bulbs in the paste-mold process.

Suck–blow process

The Owens bottle machine belongs in this category. The glass gather is sucked by the parison (preform) mold from the melt surface and is cut with automatic shears. A short plunger or core projecting into the mold produces a hollow space in the glass which is enlarged by blowing, while the parison mold is opened during reheating. For final blowing the parison, which is held by the neck mold, swings into the blow (finishing) mold (Fig. 5.6). This production method was replaced by the blow–blow process with the advent of the gob feeder.

Fig. 5.5 I-S machine production of glass containers.

Blow–blow process

The elongated gob which is formed by a gob feeder falls into the parison mold and is blown from below against the bottom of the inverted parison mold. The rest of the process is the same as the suck–blow process (Fig. 5.7).

In addition to blow–blow, the most modern development is the press–blow process. In some cases both processes can be used on the same machines with adequate tool sets.

Press–blow process

At the preform stage, the parison is not blown, but pressed by a plunger extending from the neck mold. This process permits high production rates because the parison is quickly cooled. It also gives more controlled glass distribution, allowing the manufacture of thin-walled glass containers, such as lightweight bottles.

For very narrow-neck diameters, however, not all mechanical and thermal problems are resolved yet. Both of the preceding processes yield a surface quality and wall thickness that are

90 Hollowware and glass tubing

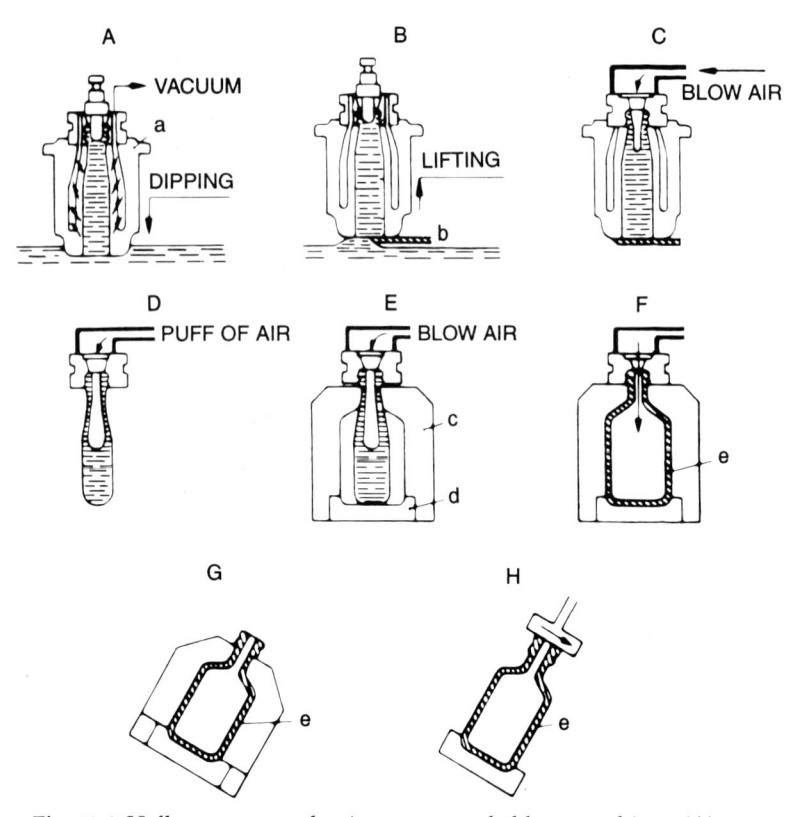

Fig. 5.6 Hollowware production on a suck–blow machine: (A) preform filling in a dipping/sucking process; (B) glass shearing; (C) attachment to air supply; (D) foreblow, reheating and stretching; (E) insertion into the finishing mold; (F) finishing blow; (G) removal from the mold and cooling; (H) detaching the finished bottle. (a) Preforming mold; (b) shears; (c) finishing mold; (d) bottom mold; (e) finished glass bottle.

not acceptable for a number of glass products. These products are reserved for the automatic paste-mold process.

Rotary–mold (paste-mold) process

The formation of the gather can be accomplished by suction or gob feeding. The lens-shaped or hollow-bodied gather, preformed by suction or by pressing of a gob, is held by a metal

Fig. 5.7 Container glass production in a blow–blow process.

ring on which a blowhead is located. Several puffs of air and the action of gravity cause the parison to lengthen until the water-saturated paste mold closes; the final blow begins with simultaneous counter rotation of the glass and the mold. The solidified glass which comes in contact with the retaining ring and the blowhead is not usable and is cut off and scrapped like the cap produced in manual production.

In addition to the blowing process, there are several other important methods for the manufacture of hollowware.

5.2.3 Pressing

In the pressing process, unlike in blowing processes, the glass-gather is in contact with all parts of the metallic mold material. As a rule, the pressing mold consists of three parts: the (hollow) mold; a plunger, which fits into the mold leaving space determining the thickness of the glass wall; and a sealing ring which guides the plunger when being removed from the mold.

A gob is fed into the mold and is hydraulically or pneumatically pressed by the ring-guided plunger until the glass is pressed into all areas of the mold. After solidification, the

Fig. 5.8 Automated glass bowl pressing.

plunger is withdrawn. Most pressing machines operate on turntables which have 4 to 20 or more molds. The turntable takes the glass step-by-step through the loading, pressing, cooling, and other processing stations to the removal point.

Typical pressed items are heat-resistant household glassware, drinking glasses, lamp globes and parts for television tubes (Fig. 5.8).

5.2.4 Extrusion

Extrusion can be used for glasses with a steep viscosity curve or glasses with an increased tendency to crystallize to produce items with close dimensional tolerances. This is an economical method of making various types of full or hollow profiles with sharp-edged cross sections for industrial use. By using laminate extrusion methods, two or three types of glass can be combined to produce, for example, components sheathed with chemically resistant glass.

5.2.5 Spinning (centrifuging)

This shaping procedure is fairly new, and not very popular. It is based on the principle that a rotating liquid will take up a surface shape of a rotation paraboloid thus forming a hollow space.

In operation, an extremely liquid glass-gather is fed into a steel mold which is then rotated at the required rate. At high revolution rates, the glass paraboloid becomes very steep and is almost cylindrical. When the glass has cooled sufficiently to solidify, the rotation is stopped, and the glass is removed. An example of a spun article is the borosilicate glass column section used in chemical processing plants. Such column sections can be made with diameters of up to 1 meter. Other non-rotationally symmetrical items, such as funnels for television tubes or other items for household and technical applications, are sometimes also made by spinning.

5.3 THE DRAWING PROCESS FOR GLASS TUBING

Glass tubing is not exactly a hollowware product, but it merits mention at this point, because a number of tubing products are converted into containers, such as ampoules, vials, fluorescent light tubes, etc. It is now rarely produced manually.

The most widely used method for a continuous drawing of tubing is named after the inventor Edward Danner, an American engineer who invented the process in 1912 (US Patent 1218598 of Libbey Glass Co.). A continuous strand of molten glass flows onto a slightly angled, slowly rotating fire clay core, called the Danner mandrel. At the lower end of the mandrel, a hollow bulb forms from which the tubing is drawn. Air is blown through the hollow mandrel shaft maintaining a hollow space in the glass. After being redirected horizontally, the solidifying tube is transported on a roller track to the pulling unit, behind which it is cut into sections of 1.5 meters length. The output of such a machine can be 3 meters per second or more (Fig. 5.9).

5.3.1 Other tube drawing processes

The Vello process is as important as the Danner process and has about the same output. The glass from the furnace forehearth flows downward through an orifice (the ring) and the hollow space in the glass is maintained by a pipe with conical opening (the bell) located within the ring. The tube, which is still soft, is redirected horizontally and is drawn off along a roller track, cooled, and cut as in the Danner process.

In the Schuller-Metz (Updraw) process, the glass tube is

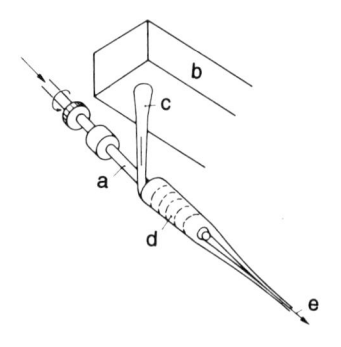

Fig. 5.9 Danner tube drawing process: (a) rotating mandrel; (b) glass feeder; (c) glass flow off the feeder; (d) glass rope wound on mandrels; (e) draw direction.

drawn vertically upward from a rotating bowl. The drawing area is shielded by a rotating ceramic cylinder, one end of which is submerged in the glass mass. The hollow space is formed by means of an air jet placed below the surface of the glass. Tubing with thick walls and large diameters can be made using this process.

In a variation of the Vello method, glass is drawn downward (Downdraw process) through a vacuum chamber. The glass passes through a sealed iris diaphragm (a circular shutter which can be adjusted for different sized openings). At a drawing speed of several meters per second, tubing with a diameter of up to 360 mm can be produced by this method.

If tubing with close tolerances is needed for special applications such as in bottle filling machines and flowmeters, this can be accomplished by shrinking the reheated tubing under partial vacuum onto a steel mandrel with the desired diameter (the Schott 'KPG' process).

Different glass types are used for tubing, depending on the application or further processing requirements: soda-lime glass is suitable for fluorescent light tubing or small lamps; highly insulating lead glass is required for electronic tubes; chemically resistant glass is required for pharmaceutical use; and borosilicate glass is used in the chemical industry.

Fig. 5.10 Glassblower with lamp.

5.4 FINISHING OF HOLLOWWARE

The processing and finishing of hollowware includes all stages necessary to produce a new item from preformed glass. This is usually done by reshaping the glass under the influence of heat. All kinds of glass tubing are used to make goods in the hollowware finishing industry.

5.4.1 Torch blowing (lampworking)

The manual process for hollowware finishing is called lampworking. The lamp is a gas torch which together with a Bunsen burner is used to heat the raw glass. The glassblower (who should not be confused with the glassmaker who works the raw, molten glass from the melting furnace) uses temperatures between 600 and 1700°C, depending upon the glass type. The tools he uses depend upon the product on which he is working; the glass can be soft or hard, referring to whether it melts at lower or higher temperatures. The glassblower's primary tools include the following: a glass cutting knife, flaring tool, carriage, nippers, forceps and measuring tools (Fig. 5.10).

Glass instruments and apparatus are typical examples of lampblown items. These include such medical and veterinary equipment as syringes, blood sampling and testing devices,

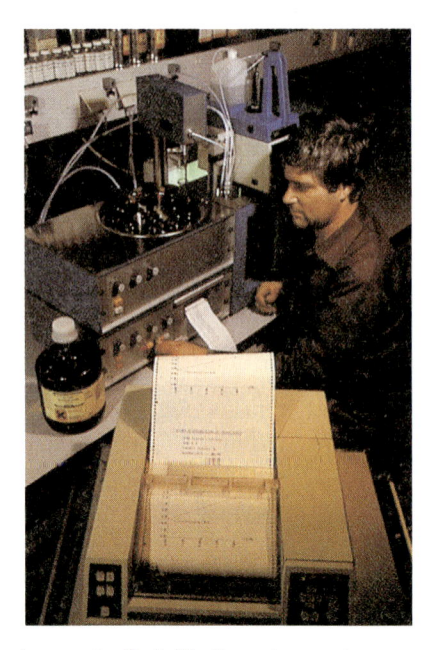

Fig. 5.11 Computer controlled dilution viscometer.

pipettes and many more. In addition, apparatus for measuring pressure and flow of gases and liquids, for viscometry, electrochemistry and water distillation can be mentioned (Fig. 5.11). In food processing plants and dairies, there are glass instruments for a myriad of uses such as acidity testers and volumetric equipment. In laboratories, one finds condensers, retorts, burettes, drop counters, separators and glass items for analysis. One important group is thermometry, including thermometers used in health care, industry, the home and in science. Lamp blowing has also enjoyed an increasing use for artistic creations (Fig. 5.12).

5.4.2 Industrial hollowware processing

As suitable packaging material for its products, the pharmaceutical industry requires large quantities of ampoules, vials, small bottles and tablet container tubes all made from glass tubing. In order to be compatible with the high-output filling machines in the pharmaceutical industry, they must be

Fig. 5.12 Lampblown art glass objects by Kurt Wallstat.

Fig. 5.13 Production of small pharmaceutical vials.

made to precise dimensions using automatic equipment (Fig. 5.13).

Simple glass parts can now also be made by semi-automated processes, especially when they are standardized. Glass funnels for TV tubes are a good example.

5.4.3 Insulating vessels

This term, normally encountered only in industry and trade, designates a useful consumer item available in nearly every household, namely, the thermos. It operates on the physical principle of heat and cold insulation by a double-walled glass container. Two glass vessels are fused in such a way that a hollow space remains between them. Another process consists of blowing a double-walled cylinder right at the furnace. Both glass bodies are later silver-coated on their interspace surfaces through a glass exhaust tube at the bottom of the vessel (Reflective flat glasses, Section 4.4.3). Finally, the air is drawn out, and the tube sealed off. The hollow, evacuated space inhibits the heat exchange between the contents of the container and the environment, and the mirrored surfaces help to prevent thermal radiation transfer. The insulating vessel is normally placed inside a protective container made of plastic, sheet iron, aluminum, stainless steel or brass. Many insulating vessels are referred to as insulating flasks, insulating jugs, food containers (with wide openings) and ice cube containers. So-called Dewar flasks are large capacity insulating vessels used for the storage of liquid gases and other laboratory applications.

5.4.4 Glass jewelry

The raw materials for glass jewelry (gemstones, medium-sized pearls and other fancy goods made from glass) are produced by placing a colored or crystal-white glass rod in a small furnace, bringing it back to the soft state, and then shaping as required using pressing tongs. The items of jewelry are completed by further cutting, grinding and polishing. Glass artisans from Gablonz in northern Bohemia are world-famous for their glass jewelry. For this reason it is sometimes referred to a Gablonzware. New forms of fashion jewelry made from glass are constantly appearing on the market. Glass beads, known as *rocailles*, are made by breaking colored tubing into small cylindrical pieces and rotating them in a heated drum. Charcoal, graphite, gypsum or other aggregates are added to prevent the pieces from fusing together.

Synthetic pearls are made by blowing small hollow spheres from thin glass tubing and filling them with mother-of-pearl colored paste.

Glass pearls are strung and made into necklaces, bracelets, rosaries and similar items. In addition to *rocailles* and hollow

Fig. 5.14 Typical glass containers.

beads there are solid beads which are made from colored rods, heated, pressed and ground as already mentioned. Polished glass pearls can be made iridescent (Iridescent glass, Section 5.8.3) by coating them with thin, transparent layers.

5.5 CONTAINER GLASS

This general term comprises all hollowware used for packaging, storage, preserving and transportation of beverages and other liquids, foods, chemicals, pharmaceuticals and cosmetics. Container glass is always manufactured in the glassworks from molten glass as opposed to vials and ampoules which are manufactured in the hollowware finishing sector (Fig. 5.14).

The importance of glass as a packaging material lies in its formability. It can be molded into custom-designed shapes identifying the brand or product it contains, such as wines and spirits. The consumer can immediately recognize the origin of certain wines from the shape of the bottle. Similarly, different types of spirits can be readily recognized by the shape of the bottle.

The unique properties of glass are even more important reasons for its use as a packaging material. Glass is odorless, impermeable, physically and chemically stable, sufficiently acid-resistant (even extremely so – Special glasses, Chapter 6 – except against hydrofluoric acid), resistant to alkalis, transparent, easy to clean and hygienic. A possible disadvantage is its weight, however, new developments have lessened this disadvantage. Glass containers have become increasingly lighter since they can be manufactured with thinner walls without affecting their mechanical stability.

Most glass containers are made from soda-lime glass. Relatively pure raw materials are used to melt colorless glass from which bottling and preserving jars, and bottles for beverages, cosmetics, cleaning materials and non-critical pharmaceutical products are made. When there are special requirements for chemical resistance, however, special glasses must be used (Pharmaceutical glass, Section 6.3).

The most frequently used colored glass containers are green or brown. Green glass, which transmits ultraviolet light, is obtained by adding chromium oxide to the batch. Brown glass, which absorbs almost all ultraviolet light, is obtained by adding iron sulfide, either directly as pyrites or by simultaneous addition of Na_2SO_4 and carbon (as reducing agent) which in the presence of iron produces so-called carbon amber (for other colors, see Raw materials, Section 2.4). Color can be used for identification, for protection from light, or it can be used purely for aesthetic reasons. The amount of coloring agents in a glass batch is usually less than 0.5%; and practically none are leached out by the contents of the bottle, even during long periods of storage.

5.5.1 Beverage bottles

These are made by either the blow–blow or press–blow process. In Germany bottle sizes and volumes for various beverages are standardized by law. This protects the consumer against deceit and fraud. For example, wine and liquor bottles are full bottles when their contents contain 0.7 liters. Most recently, they may also contain 0.75 liters, and are identified as such by a ring of stars. A half bottle, on the other hand, contains 0.35 liters. Beer is sold in 0.5 and 0.33 liter bottles. Other bottled products have other standard volumes. Bottles having contents of 1, 1.5, 2, 3 and additional full liter amounts are allowed for all products.

Legislators have determined in which cases a bottle may serve as a unit of measure. The precision of bottle dimensions is very important here, and the National Institute of Standards and Technology closely controls this by taking random samples in glass factories. Bottles marked with an M on the bottom are legally certified as units of measure in accordance with this legislation. Also stamped on the bottom are an official manufacturer's trademark and a rated volume. Thus beverage containers, unlike bar glasses for beer, spirits, etc., do not need a calibration mark.

Glass balloons or carboys can contain up to 60 liters. They are usually wrapped in wicker or straw and placed in metallic or wooden baskets (sometimes referred to as basket bottles). Acids, alkalis, and other liquid chemicals are stored and transported in them.

In order to be competitive with lighter container materials, such as tinplate or plastic, manufacturers have long wished to reduce the weight of glass containers by making the walls thinner without affecting their mechanical strength. This has been achieved on the one hand by means of reduced but more uniform wall thickness, and on the other hand by reducing friction and minute surface fissures through hot vapor deposition of tin oxide, in combination with a plastic coating on the outside surface. This process has increased the bursting pressure by two to three times.

Owing to economic considerations, the one-way bottle was promoted as the best alternative to returnable bottles. An increased level of environmental awareness has led to the recycling system, established by glass companies in 1975. Discarded bottles are added as cullet to the batch and remelted. This results in considerable savings in raw materials and energy. One difficulty associated with this program is the necessity of sorting the discarded bottles according to color, as both clear and brown glass colorings are sensitive to impurities. Green glass, on the other hand, can accept other glass types without noticeable influence on the color.

5.5.2 Bottling jars

Industrial and household preserving vessels (bottling jars) are usually made of clear glass. Clear vision of the contents of a jar is a good sales aid, especially for foods.

The chemical resistance of such glass is very important. The resistance to aqueous solutions corresponds to hydrolytic class 3 in DIN 12111, which means that it is high enough to withstand repeated boiling with its contents and that washing processes result in no noticeable changes in the surface of the glass such as clouding or staining. Owing to the relatively high thermal expansion of soda-lime glass ($\alpha \approx 9 \times 10^{-6}$/K), rapid temperature changes must be avoided, especially in jars with thick walls or bottoms, as this can cause breakage.

5.6 GLASS TABLEWARE

This general and often confusing term comprises a group of hollowware products which are encountered in daily life and which have high demands placed upon them, especially from the design point of view. This glass is also referred to as household glassware.

The main categories of the tableware family of products are drinking glasses, which comprise, in terms of value, about 60% of all the tableware produced. The remaining 40% is accounted for by other glass accessories for the table and by articles which are used in the kitchen, home and office.

Drinking glass sets include all glasses used in a household or commercially for individual drinks. Since these often have stems, they are usually called stemware. The term 'set' denotes that they have basic designs (shapes or styles of decoration) common to all glasses belonging to that set (Fig. 5.15).

Other typical tableware products, sometimes also called giftware, include ashtrays and smoking sets, table and floor vases, flower bowls, candle holders and large glass dishes. Decorative glass items without specific functions, figurines, glass animals and wall decorations made of glass also belong to this broad group. Simple drinking glasses for catering and daily domestic use are almost always mass-produced, machine-made articles. It is also possible to machine produce stemware, including crystalware and lead crystalware,' which is distin-

Fig. 5.15 The 'Sélection' stemware set from Schott-Zwiesel.

guishable from the hand-made product only by the expert. Such products are not meant to replace manual production, which for intricate and individual designs is second to none. Instead, they ensure that a wide segment of consumers can buy high quality, mass-produced products at a reasonable price.

5.6.1 Breakdown of tableware by glass type

There are three main groups of tableware, categorized according to their composition. By far the largest group is soda-lime glass, although each of the main producers may have slightly different variations of this glass. The physical and chemical behavior of this glass is covered in the previous section.

The second group includes the crystal and lead crystal glasses, in which most of the calcium is replaced by barium, zinc or lead, while sodium is in part replaced by potassium. The increased refractive index resulting from this gives the glass a high degree of brilliance which can be further enhanced by cutting. The term crystal glass was adopted due to its resemblance to precious stones. For several years the European Community countries have had legislation stipulating the following quality criteria for crystal and lead crystal glass.

1. Crystal glass must contain a minimum of 10% lead oxide (PbO), barium oxide (BaO), potassium oxide (K_2O), or zinc oxide (ZnO), alone or in combination, and its density (d) must be at least 2.45 g/cm^3. Its refractive index (n_D) must be

at least 1.520 based on the yellow sodium-D spectral line. If there is no ZnO, however, the density must be at least 2.40 g/cm^3 and the Vickers surface hardness must be 550 ± 20.
2. Pressed lead crystal contains at least 18% PbO, $d \geqslant 2.70$ g/cm^3, and $n_D \geqslant 1.520$. This designation is approved only in Germany.
3. Lead crystal contains a minimum 24% PbO, $d \geqslant 2.90$, and $n_D \geqslant 1.545$.
4. Full lead crystal contains at least 30% PbO, $d \geqslant 3.00$, and $n_D \geqslant 1.545$.

Density and refractive index were chosen as test criteria because the glass does not have to be broken in the laboratory in order to measure them. Naturally the valuation of these glasses is not dependent upon their composition alone, but it also depends to a great extent on the finishing work involved and the standard of design. Lead crystal is generally considered to be especially valuable, however. In addition, as long glasses (General characteristics, Section 2.2), the lead glasses are easy to shape and cut, and also can be acid polished (Finishing, Section 5.8). While their hydrolytic resistance is only slightly higher than that of soda-lime glass, they have an excellent dishwasher cleaning resistance. Their thermal coefficient of expansion lies between 7.5 and 9.0×10^{-6}/K.

The third group includes machine-pressed or machine-blown tableware made of special glasses and processed directly from the glass melting tank. Their compositions vary greatly from those previously mentioned. There are currently three popular types on the market described below.

Transparent borosilicate glass

The *Duran* (Schott) and *Pyrex* (Corning, USA) borosilicate glasses which were introduced many years ago, are further developments of Otto Schott's *Geräteglas* (apparatus glass), which was introduced into households as *Jenaer Glas* during the 1920s. Dishes, baking molds, teapots, glasses and cups, coffee-maker jugs, baby feeding bottles, etc. are well-known products mass-produced at the melting tank from this type of glass (Figs 5.16 and 5.17). Thin-walled glasses are blown in hinged molds

Fig. 5.16 Borosilicate glass kitchenware.

Fig. 5.17 Pressed borosilicate glass bowls.

in a paste-mold process; thick-walled articles are produced in I-S machines or pressed. In the latter process, the hot glass is quenched with cold air in order to produce compressive stress in the glass surface (thermal toughening, see Safety glass, Section 4.4.5). The firm's logo or trademark is either pressed into the bottom, or it is silk-screened and then fired in a further processing step. Since the borosilicate glasses play an important role in technology, owing to their chemical durability and low thermal expansion, they are discussed in more detail under 'Special glasses' in Chapter 6.

Fig. 5.18 Decorated opal glass.

White opaque glass (translucent glass, opal glass)

Opaque (not transparent) glass is identified by an inhomogeneous composition containing a mixture of at least two different components (phases) in a microscopically fine distribution. Such glasses are formed when either one phase (such as tin oxide) does not dissolve during the melting process or one crystallizes out during cooling (fluorides); or else the melt separates into two phases which, although glassy, exhibit strong light dispersing effects due to their different refractive indices (as opposed to opaque tinted flat glass – see Opaque glass, Section 4.1.1 – which absorbs light). These glasses appear white or slightly tinted, depending on their composition (Fig. 5.18). The use of fluorides, which partially volatilize into the atmosphere, is now rarely practised due to environment protection laws. More recently, opacifying agents have largely been replaced by phosphates (primarily aluminum phosphate in combination with barium carbonate). The resulting white colored opal glasses, which are usually color-decorated at tableware, have a relatively low thermal expansion ($\alpha \approx 4.5 \times 10^{-6}$/K).

Glass-ceramics

The principles of the manufacture of glass-ceramics are covered extensively in Chapter 6 under 'Glass-ceramics'. The products in this group, which were first brought out under the names *Pyroceram* (opaque) by Corning and *Jena 2000* (transparent) by

Fig. 5.19 Modern cooking surface with a Schott CERAN® glass-ceramic smooth-top.

Schott, have coefficients of thermal expansion which are close to zero, and therefore, can withstand extreme temperature shocks. For example, they can be taken out of the freezer and immediately placed on a hot stove.

In its optical properties, the opaque white glass-ceramic is closely related to porcelain and is a result of the fast growth of microcrystals during the ceramizing process.

While bowls and dishes in this material are normally pressed, the same material, with a dark tint, is used to produce (in a continuous rolling process) the base material for CERAN® cooking surfaces for electric, gas and induction ranges (Fig. 5.19). Pots and pans can be moved from the heating zone into a cold stand-by zone simply by pushing them over the flat, easy-to-clean surface. The *'Ceran'* glass-ceramic cooking surfaces from Schott came into the market as an essential part of the *'Ceran-top-system'* in the early 1980s and have since changed the cooking habits and the design of kitchens in Europe and the USA.

5.7 OTHER HOLLOWWARE

The technical applications of hollowware are so numerous that it is impossible to describe them all. So only some of the most common uses will be discussed here; others will be mentioned in Chapter 6.

5.7.1 Hollow structural glass

Glass blocks, 'concrete' glass and glass roofing tiles comprise the architectural hollow glass sector. All of these glasses are manufactured by pressing. Rectangular glass blocks are made by fusing two pressed halves together, whereby the pressure of the enclosed air is greatly reduced through the cooling process. This creates good thermal insulating characteristics and sound absorption (up to 40 decibels and more). The exposed surfaces can also be made with ornamental designs or light-scattering textures, colored decorations, or solar protective coatings. There are also solid glass elements in the form of panels and tiles. With appropriately fitting profiles and cementing materials, the individual building blocks are assembled into walls or window panels. They transmit light but do not allow undistorted vision. They are not approved for use in load-bearing applications.

Concrete glass, which is also pressed into solid or hollow building blocks, is used to manufacture glass-crete, a material made of glass, steel, and concrete. This is used for light transmitting floors and light shafts to be walked on and, in some cases, even driven on. Glass roofing tiles are shaped like common clay tiles and must be strong enough to be walked on and to be hailstone-resistant. Concrete glass windows of a purely decorative nature have already been discussed in Chapter 4.

5.7.2 Lighting glass

This glass should not be confused with lamp glass (Lamp glasses, Section 6.4.9) which is made mainly from special glasses. Lighting glass is used primarily for all kinds of lighting fixtures, including projector lamps. Besides hollowware parts, a considerable amount of pressed glass parts is being used for this purpose. Like tableware, most lighting glass is machine-made soda-lime glass. For higher brilliance crystal and lead crystal is used (in this application the crystal glass labelling rules do not apply). Also, opal glass as well as borosilicate glass is used for high-powered lights (projector lamps, copying machines, oil lamps). Another variation of glass used for lampshades is flashed opal glass (Flashing, Section 5.8.1). The manufacture of flashed opal glass has already been described in Chapter 4. The high degree of light diffusion of these glare-free glasses results in a certain amount of light loss; but it provides even, glare-free illumination.

Colored glass is also flashed for lighting applications. Custom designed lighting glass is supplied to the lamp manufacturers for the electrical hardware equipment and final assembly. There are still glass factories that also produce standard articles which they make into finished lights.

5.7.3 Laboratory glass and medical hollowware

This category comprises a large group of glass products which are used in many different areas, ranging from the laboratory to industry, universities, standards bureaus, clinics and chemical process plants. A large and still growing portion is made from special glasses, covered in Chapter 6.

5.8 FINISHING OF HOLLOWWARE

Strictly speaking, the finishing of hollowware relates to the further processing of the outer surface of the basically finished glass as it comes directly from the glass plant. This is similar to what is described in the flat glass section.

In reality, the term is used much more broadly. It is divided into two categories: finishing in the hot state at the furnace and finishing in the cold state. Finishing at the furnace can only be done in the glass factory following the basic shaping or molding. Cold finishing is done in further processing steps in the factory after the glass has been formed or in finishing plants which purchase the raw glass article from a glass manufacturer. Sometimes even cold finishing requires a reheating of the glass.

5.8.1 Finishing in the hot state

The techniques involved here are of prime importance for tableware and lighting glass. Not all of them are significant for the economy; many have only historical importance or are no longer economical. It is worthwhile to present these older methods of glass finishing not only for nostalgic or historic reasons, but also because some of these methods may even be further developed using today's advanced knowledge.

Optically blown glass

The light refraction of glass can be enhanced by a rippled surface. To do this, the small parison is blown in a cylindrical, corrugated hollow mold. The surface texture is maintained even during the blowing process, and expands with the growing size of the finished article. This is how longitudinally striped cylinder lenses are made. If they are twisted, they form a spiral.

Application of filaments and drops

Molten glass drawn into thin filaments is arranged in specific designs onto the hot finished glass article. Depending on the arrangement of the filaments, this results in a spiral, ring, free-form, or picture design (Fig. 5.20). Instead of filaments, individual droplets of glass called nubs can be attached to the hot surface. Pointed nubs hanging downward are called snouts.

Flashing

Dipping a colorless glass parison into colored glass, followed by blowing, produces a so-called external flash. If the process begins with a colored parison, it is called an internal flash (Fig. 5.21). A combination of both results in a double flash.

Fig. 5.20 Vase with filament application.

Fig. 5.21 Pitcher with olive colored internal flash and gold-brown ribbon in bubble technique.

Subsequent grinding or etching through the outer layer results in interesting effects caused by the contrasting colors.

Mosaic glass

Pieces of glass of various colors are spread over a marble or iron plate. The glassblower then rolls the hot glass globe, in

Fig. 5.22 *Millefiori* bottle.

clear or colored glass, over it and the particles of glass immediately adhere to the surface. Then the entire object is overlaid with clear glass and finished in the usual way.

Millefiori glass (thousand flower glass)

Multicolored glass rods are bundled together, fused, and sliced into thin pieces. Further processing is similar to that of mosaic glass (Fig. 5.22).

Filament inlay

A viscous filament of glass is laid on top of the hot glass surface and pressed into the surface by rolling back and forth on a marble or iron plate. Comb-like tools can also be used to line up several filaments at the same time, resulting in a 'combed' design. The color of the filaments must be different from the base glass; otherwise, they would not be visible. There is another process in which thin rods, usually made of opal glass, are arranged along with colorless rods on the inner wall of a clay tube, and clear glass is then blown from inside against these rods. The rods fuse with the main body of glass and result in a vertically striped piece of blown glassware, which can be made into a spiral by reheating and twisting. The

Finishing of hollowware

Fig. 5.23 Snuff flasks with filament inlay.

entire object is then covered with a layer of clear glass (Fig. 5.23).

Air bubbles

If the still-soft glass surface is pricked with needles or nails, a hollow space remains inside the glass after they are removed, while the surface seals up again. Each puncture leaves a bubble in the glass. The same effect can be achieved with molds with needles inside. Hollow stems in stemware are produced in a similar manner. They can be several centimeters long and can also be twisted.

Crackle glass (craquelé)

The parison is rolled in moist sawdust or covered with coarse sand to intentionally damage the surface. Then it is dipped in water. The quenching causes just the surface to crack without destroying the glass. The blown glass is then covered with a fresh layer of glass and reheated until the cracks fuse slightly together so that the glass maintains its stability (Fig. 5.20).

Fig. 5.24 Marbled vase (agate glass).

Foundry ice

The hot glass body is rolled over grainy glass powder, or the powder is sprinkled on it. A more or less rough surface results, depending on the size of the grains used.

Agate glass

A marble-like glass structure can be created by mixing molten glasses of different colors. The result is a striped pattern named after the semi-precious gemstone which it resembles (Fig. 5.24).

Flame coating with metals

Glass forms a solid bond with metals (Cu, Al) in the hot state (above 250°C). To a certain extent, the glass and the metal fuse together. The higher the temperature of the glass, the better the adhesion.

A flame gun is used to spray the metal on the hot glass (usually done as the glass comes out of the furnace to save energy; otherwise, the glass must be annealed and then reheated). In this process, a metal wire or metallic powder passes through the center of an oxyacetylene flame and is melted by the high flame temperature. It is virtually 'shot' onto the surface of the glass, where it bonds chemically to the glass on contact. The surface becomes metallic. It can be altered or

Fig. 5.25 Cutting of a pattern into crystal glass.

finished by brushing or buffing. This kind of glass is used for lighting, double-glazing and other applications.

5.8.2 Finishing in the cold state – glass removing processes

After the blown hollowware has passed through the annealing lehr, and after completion of such other hot finishing processes as removing the blowing cap (excess glass) in tableware and fire polishing the edges, it can be further finished by various techniques.

Cutting

This is the most common process used to finish or decorate hollowware articles (Fig. 5.25). Horizontal and vertical grinding wheels are used in this process. If the grinding surface of the wheel is flat, flat or thumbnail grinding results (Fig. 5.26). V-shaped grinding wheel surfaces produce wedge cuts. If the wheel is rounded in a convex shape, the result is called hollow grinding. With spherical wheels the cutter can produce circular cuts, called spheres, or elliptical ones, called olive cuts.

Cutting patterns are composed of several spherical or otherwise shaped areas. Wedge cutting can also be added. Com-

Fig. 5.26 Crystal glasses with thumbnail cut (left), diamond cut (center), notch cut (right).

binations of cutting styles allow different decorative motifs to be cut into the glass, such as flowers. Usually the cutting is done in two or three steps which are performed by different workers. First, the pattern is rough-ground into the glass using an iron wheel and abrasives (such as sand, carborundum (SiC)) which remove most of the glass. Finer cutting then follows. In addition to these manual processes, cutting machinery has been in use for a number of years.

Polishing

When the glasses leave the grinding shop, the decorative cuts are still matt ground. In order to achieve the required high degree of brilliance, one of three additional processes can be used.

Acid polishing is done by placing the glasses in wire baskets, then dipping them in a mixture of hydrofluoric acid, sulfuric acid, and hot water, followed by a clean water rinse. This process lasts up to 30 seconds, and is repeated several times. Large batches of decorated glasses can be polished quickly with this process.

Mechanical polishing is usually done with poplar wood wheels using a polishing agent (usually polishing rouge, a fine-powdered iron oxide).

Lastly, there is fire polishing. Here the glass is heated to 500–700°C, whereby the softened, viscous glass surface shrinks and smoothens through the surface tension effect.

Engraving

This is a refined method of glass-cutting. Small copper wheels with diameters of 2–100 mm rotate on a horizontal axis. A grinding agent (emery and linseed oil) drips on the wheels and removes glass from the glass article when it is held against the wheel from beneath. In raised or cameo cutting, the surrounding glass surface is removed and a raised picture remains. The more popular deep cut leaves the glass surface intact and cuts the design into it. Line engraving is done like deep cutting, however, it results in only fine-lined designs.

Etching

This is based on the principle of acidic corrosion of glass. Hydrofluoric acid is most often used. Sometimes hydrofluoric acid vapors or baths of hydrofluoric acid salts are used. This depends on whether the intended result is a deep etching in the glass or just a frosted surface. The expert speaks of deep baths and matt baths. The matting (or frosting) of finished glass articles is important in lighting glass. Flashed opal glasses can be smooth or frosted, for example. Because of the danger associated with hydrofluoric acid vapors and splashes, this etching process is avoided whenever possible. In many cases, it can be replaced by sand blasting.

Etched designs are created by inscribing patterns into a wax coated glass, either with a pantograph or with a geometrical lathe which can batch-process 24 or more glasses. When etching acid comes in contact with the object, the wax covered portions of the glass are not attacked. Patterns and marks can also be applied directly onto the surface by using etching ink with either gold or platinum pens, or rubber stamps. This is how calibration marks are applied on many glass items.

Sand blasting

This method is used to create not only matt surfaces but also deeper cuttings in the glass. Glass surfaces that should not be attacked are protected from the sand or corundum grit (Al_2O_3) by lacquer coatings or small rubber masks. As in the etching process, the covered surfaces of the glass remain clearly transparent. Decorative patterns formed by sand blasting are well-suited for vases, large bowls, and other types of art glass.

5.8.3 Surface coating processes

Painting

Enamel colors, which are fired in at 550–650°C, are used for this process. Basically, they are nothing more than low melting pulverized colored glasses mixed with liquid (such as turpentine).

Such colors are called enamels and are either opaque (containing opacifiers) or translucent. Drinking glasses, pitchers, vases and lamp bases can be decorated by painting. Black lead is also an enamel, consisting of lead glass with additions of iron and copper. It was widely used in medieval glass paintings.

Color staining

Color staining is done mainly with compounds of copper and silver. They are mixed with kaolin or ochre, stirred into a liquid, and applied to cold glass. Upon heating the glass to almost the transformation temperature, an ion exchange takes place between the alkali in the glass and the silver or copper ions. There are ruby stains and yellow stains. Black tints can be produced through migration of silver caused by hot reducing gases such as hydrogen. Stained hollowware, such as vases or goblets, can then be engraved or cut in an additional process.

Painting with precious metals

Silver, gold, or platinum mixed with certain substances can be painted on and fired into glass. After firing, the metal can then be polished to its natural luster. This technique is most often performed on the rim of expensive drinking glasses. If an etched design is first applied to the glass, and silver, gold or platinum is brushed over it and polished after firing, the cavities of the etched design remain matt.

Rubber stamps

They are used to create single color decors. The color is then fired in.

Steel stamping

Mainly single color designs or drawings are manually or mechanically engraved in a steel plate or applied photographically. After the plate is colored, the picture is transferred onto pliable paper which is then pressed onto the glass. The design is then fired in.

Iridescent glass

Shimmering colors can be obtained when the glass is heated to at least 400°C, and vapors of solutions (e.g. chlorides) containing multivalent metals combined with moist carrier gases are directed over the glass surface. The compounds decompose on the molten glass and are broken down into oxides. The colors appearing mainly in reflected light result from optical interference of the light waves when the thickness of the oxide layer is at least one quarter of the color's wavelength and its refractive index is substantially higher than that of the glass. The oxides of tin, titanium and bismuth are used, among others. As mentioned previously, similar effects are also obtained by using a vacuum-depositing technique in the production of glass jewelry.

Fig. 5.27 Example of decal decor.

Decals

Multicolored picture negatives or letters are first printed on a special paper substrate. They are then covered with another protective layer of paper. In order to apply the decal, the protective paper is removed and the paper substrate is pressed onto the glass. After rinsing with water, the paper substrate can be removed. The transfer picture remains on the glass, and the colors are fired in (Fig. 5.27). Promotional glasses, for example, which are usually ordered in small quantities, can be decorated this way.

Silk screening

This is a modern process for decoration in mass production. Photographically produced screens are used in the printing process, with a separate screen being used for each color. The screening machines operate with great precision so that even when several separate color printings are necessary, the register of each color is exact. After screening, the glasses pass through a firing furnace. The logos and names on beer glasses, and brand names on beverage bottles and jars for preserved goods are usually produced using a silk screen process.

6
Special glasses and their uses

Unlike flat glass, glass fibers or hollowware, special glass is not identified by its appearance. The most significant factor for the various types of special glasses is their application and this determines the requirements on certain properties of the glass. These glass properties can be influenced or adjusted by formulating suitable compositions which, in turn, requires intensive scientific research which few glass companies in the world can support. The result is a range of special glasses with high chemical and thermal durability and with a variety of optical, electrochemical or special application-tailored properties. These glasses are used in many different fields, such as chemistry, pharmacy, electrotechnology, electronics, apparatus and instrument construction, optics, illumination engineering, household appliances, certain sectors of the construction industry and in other technical applications.

6.1 FUSED SILICA (FUSED QUARTZ OR QUARTZ GLASS)

Of the single-component glasses, only the SiO_2 glass (quartz glass or fused silica) has technical importance. This is primarily because of its very low thermal expansion ($\alpha \approx 0.5 \times 10^{-6}/K$), its high thermal stability (up to almost 1000°C), and its extremely high ultraviolet transmission (Table 6.3, Glass No. 1). Because the glass must be melted at temperatures above 2000°C, it is expensive and difficult to produce. A less expensive substitute is vitreous fused silica, which is a quartz glass melted at lower temperatures. Since it is not thoroughly refined, it is laced with small bubbles and is not transparent.

Another process for making clear quartz glass starts with boron-enriched, phase-separable alkali-silicate glasses with low

122 Special glasses and their uses

Table 6.1 Physical and chemical properties of *Duran* glass (Schott)

Density	2.23 g/cm^3
Linear thermal expansion	3.2–3.3 × 10^{-6}/K
Modulus of eleasticity	6.3 × 10^4 N/mm^2
Tensile strength (fire polished)	~90 N/mm^2
Calculated continuous load factor	6 N/mm^2
Refractive index (n_D)	1.473
Electrical volume resistance at 250°C	10^8 Ω cm
Transformation temperature (T_g)	530°C
Softening point	815°C
Chemical resistance to attack by	water, according to DIN-ISO/719 (5 classes) : 1 acids, according to DIN 12116 (4 classes) : 1 alkaline solutions, according
(Best resistance is class 1.)	to DIN 52322 (3 classes) : 2

melting points. When the glass is thermally treated at around 600°C it separates into two phases. The alkali-borate rich phase may be leached out with acids, leaving behind open pores of controllable microscopic sizes.

In a subsequent heating process, the remaining high silica (approximately 96%) phase can be transformed into a clear glass. Known as 'Vycor', it has properties similar to pure fused silica (Glass No. 2 in Table 6.3). For optical components such as ultraviolet light guides (Glass fiber optics, Section 6.7.3), high-purity fused silica is obtained by pyrolytic decomposition of gaseous silicon halogen compounds ($SiCl_4$).

By adding 7–10% by weight of titanium oxide to the SiO_2, the thermal expansion can be lowered to slightly negative values. Glasses with this composition were used in the USA as mirror substrates for telescopes. More recently, however, they have been replaced by glass-ceramics (Glass-ceramics, Section 6.8).

Leached SiO_2 glasses of the Vycor type with defined pore sizes are used as membranes in ultrafiltration and dialysis (e.g., for separating oil emulsions in water), and as carriers for biologically active materials, such as enzymes, in the food processing industry.

6.2 BOROSILICATE GLASSES FOR INDUSTRIAL AND LABORATORY USE

The widespread group of borosilicate glasses has been mentioned on several occasions. A broad description of the composition and properties of the main types of borosilicate glasses was presented in Chapter 2, and their importance in household and consumer glassware was mentioned in Chapter 5. The following sections will discuss the purposes which these glasses serve in many technical areas – the chemical, pharmaceutical and electrotechnical industries in particular. Technical data and the most important application-oriented properties of the main representative of this group, the Schott glass *Duran*, are listed in Table 6.1. The American borosilicate glass, *Pyrex*, shows an almost identical behavior.

In addition to being almost insensitive to temperature shock, borosilicate glasses exhibit other advantageous qualities: they will not deform below 550°C; they have no gel layer formation on the surface (hence low resistance to the flow of liquids and little tendency to catch solid depositions); they release no metals when in contact with liquids; they do not act as a catalyst; and they are resistant to radioactive radiation (Fig. 6.1). As long glasses (General characteristics, Section 2.2), they are suitable for hot processing even for complicated component shapes and apparatus.

6.2.1 Laboratory equipment

Only a few characteristic examples shall be cited here out of the broad variety of borosilicate glass apparatus used in research, development and test laboratories of all kinds, and in process technology. Beginning with test tubes, standard laboratory glassware includes beakers and flasks, volumetric glassware (such as graduated measuring cylinders, pipettes, and burettes), filtering devices, gas washing bottles and reaction vessels, distillation equipment, condensers and heat exchangers, as well as stopcocks and valves for liquids and gases. Special glass types are used to make thermometers; alkali-free glass enables precise temperature readings of up to 620°C. According to the type of laboratory, additional specialized apparatus and measuring equipment made of glass permits testing, conver-

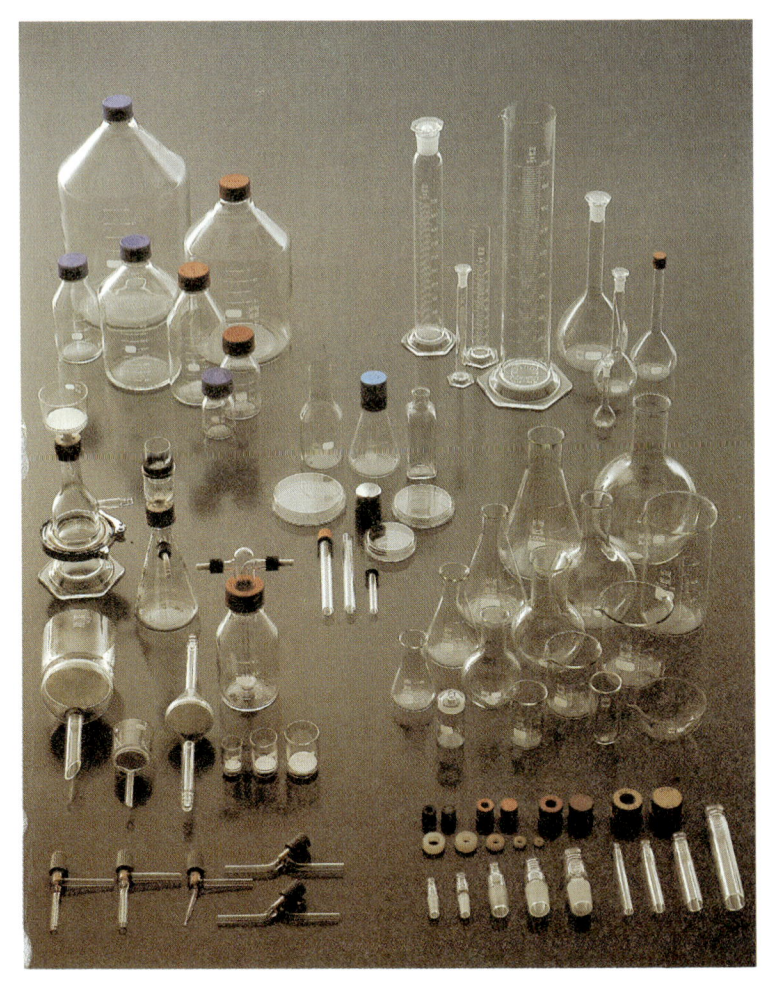

Fig. 6.1 Laboratory glass.

sion or separation of materials. Examples include apparatus for gas analysis, titration, liquid and gas flowmeters, capillary viscometers for viscosity measurement by flowrate, and equipment for molecular distillation used in the production of vitamin extracts or essential oils and fragrances.

Accessory process equipment, such as stirrers, pumps, high-vacuum and diffusion pumps, and rotary evaporators also

Fig. 6.2 Centrifugal pumps for chemicals.

belong to this product group (Fig. 6.2); along with different types of quick-fitting connectors for processing equipment components, which utilize tapered and flat-range joints or glass threads, assuring liquid- and gas-tight seals in all cases.

6.2.2 Glass process plant

This is a large scale extension of laboratory equipment; the difference is that the amount of material to be treated or processed is many times larger. Such processing equipment in glass is generally built as a pilot plant before being constructed for full-scale production.

By using standard interchangeable parts, it is possible to construct complicated facilities having many components, large dimensions, long pipelines, and many side branches. Currently, borosilicate glass can be shaped into pipes having diameters up to 300 mm and vessels with 500 liter useful capacity (Fig. 6.3). Straight pipes and bends, valves, column sections and heat exchangers are available in a range of nominal widths and modular lengths. Various types of glass pumps working on different principles are also used in such constructions.

Glass conduits of any length for wastewater or gas drainage can be installed inside and outside buildings in pipe ducts or chases, or even underground. To avoid excess strain, they are not installed rigidly, but mounted in suspension straps and pipe clamps. For pipes with thin flanges at each end, it is

126 Special glasses and their uses

Fig. 6.3 Tubular borosilicate glass heat exchanger.

sufficient in many cases to use rubber sleeves and metal clamps.

In addition to spherical and cylindrical vessels, columns in sizes ranging up to 20 meters in height and 1 meter in diameter are used in chemical process systems (Fig. 6.4). They are used for extraction and heat exchange between rising vapors and descending liquids. They usually also contain small rings of glass tubing (Raschig rings) which serve to increase both the surface area for and the time of contact between the flowing media. The throughput can be as much as several tons per hour. The gaskets used between the rigid glass components are mostly made of flexible PTFE rings with a temperature resistance of up to 200°C. Depending on the nominal width, they resist a pressure of between 0.5 and 4 bar.

Processing equipment in glass is used in the pharmaceutical industry for extraction and distillation of substances and for solvent recovery. Distilleries use continuous rectifying columns to purify and concentrate distillates. Glass pipelines are widely used in all types of beverage production, in dairies, in the production of aromatic compounds and in special branches of the food processing industry. Textile finishing plants use them in dyeing equipment. In the electroplating industries, vacuum concentrators made of glass are used to recover metals and chemicals from the electroplating baths. The dimensions of plant and pipeline components are extensively standardized (e.g., ISO and DIN) and these standards are constantly revised

Fig. 6.4 Chemical process plant made of borosilicate glass.

and updated. Components of different manufacture, therefore, can often be used together.

6.3 PHARMACEUTICAL GLASS

Many pharmaceutical preparations and drugs are packaged in glass containers such as ampoules, vials, and small bottles (Fig. 6.5). While ampoules are always made from glass tubing, small bottles can be made either from tubing or directly off the melting tank. The pharmacopoeia* (European = Eur. AB; German = DAB) dictate that injectable substances may only be bottled in containers, the hydrolytic surface resistance of which meets the requirements of Glass Type I of the Eur. AB (Durability Class I) or Durability Group A of the DAB (in

---

* 'Pharmacopoeia' are official collections of regulations concerning quality, dosage, testing, and storing of drugs. Most advanced countries publish them.

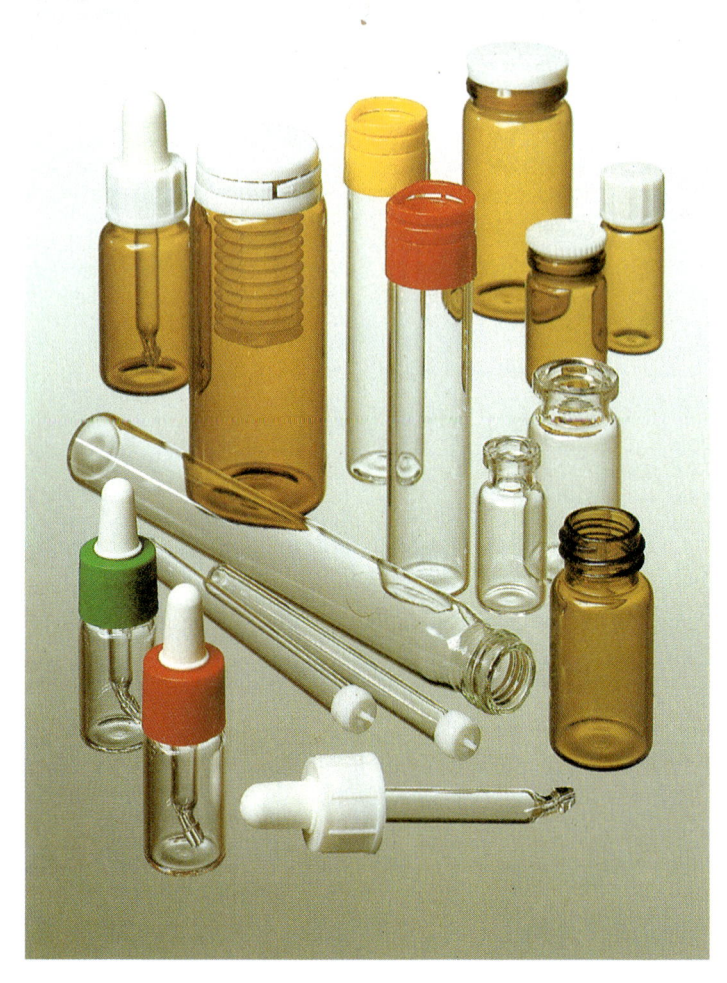

Fig. 6.5 Pharmaceutical vials.

accordance with DIN 52329; or ISO 4802 in preparation). The *Fiolax* clear (also called *Fiolax* colorless or white) and the *Fiolax* brown made by Schott Rohrglas are borosilicate glasses which meet these requirements. The glasses contain less boric acid than *Duran* glass (see above) and more alkali oxides, as well as a considerable percentage of calcium oxide and barium oxide.

Table 6.2 Chemical and physical properties of glass tubing for pharmaceutical applications

	Fiolax Clear	Fiolax Brown	Illax	AR-glass
Hydrolytic class	1	1	2	3
Surface resistance (Eur. AB)	I	I	I	III
Surface resistance (DAB)	A	A	B	C
Acid resistance	1	1	2	1
Resistance to alkaline solutions	2	2	2	2
Coefficient of thermal expansion ($10^{-6}\,K^{-1}$)	4.9	5.4	7.6	9.0
Transformation temperature, T_g (°C)	560	550	528	520
Working temperature, V_a (°C)	1160	1145	1058	1035
Density (g/cm^3)	2.39	2.44	2.50	2.52

The coloring agents for the brown glass are iron oxide and titanium oxide. Both glasses are well-suited for pharmaceutical use due to their high chemical durability (see Table 6.2). In addition, they possess good glass processing characteristics and are highly suited for automatic machinery processing of glass tubing into containers with the close dimensional tolerances required by the very fast filling equipment used in the pharmaceutical industry.

The spectral transmission of the brown *Fiolax* glass, which is used to protect contents against sunlight, is shown in curve (a) in Fig. 6.6. Curve (b) represents *Illax* glass which is colored with iron oxide and manganese oxide and used for less sensitive preparations. It has a high degree of chemical durability, but it must not be used for injectable substances because it is a class 2 glass in regard to hydrolytic resistance. It is used rather for drinking vials, containers for sensitive reagents, tubes for pills which are sensitive to light and air, medicine bottles, etc.

The last glass in this group is the AR-glass which is a colorless soda-lime glass containing only 1.5% boric acid. Since this glass is in hydrolytic class 3, it is mainly used for bottles for dry or water free (oily) preparations.

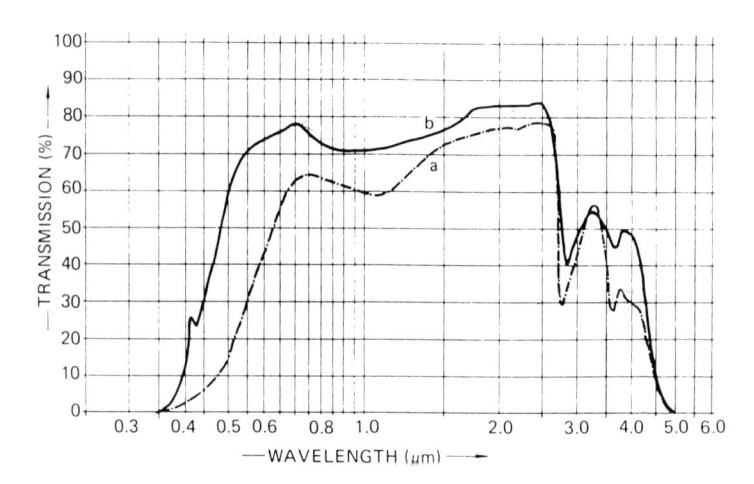

Fig. 6.6 Spectral transmittance of (a) *Fiolax* amber and (b) *Illax* at 1 mm thickness.

6.4 GLASSES FOR ELECTROTECHNOLOGY AND ELECTRONICS

The special glasses used in electrical engineering are characterized by excellent electrical insulation, low dielectric loss, gaseous impermeability, high absorption of certain radiations, and 'tailor-made' thermal properties. The most important requirement comes from the vacuum tube and semiconductor technology where electrical conductors have to be fed through glass bulbs or metal housings in an absolutely hermetic glass-to-metal seal. Various conductor materials and alloys are used. These seals require painstaking matching of the coefficients of thermal expansion (α) of the glass and metal over a wide temperature range to ensure hermeticity and freedom from microcracks under all possible operating conditions. Such glasses are called sealing glasses.

In many cases, additional special properties are required, e.g., resistance to certain gases and vapors (in alkali vapor lamps), absence of alkali oxides, and low working temperatures.

6.4.1 Sealing glasses

The various glasses for glass-to-metal seals are grouped according to their match to the thermal expansion ranges of the available sealing metals. When a direct seal between a glass and another material is not possible because of residual stress created by large differences in α-values, one or several transition glasses must be used between the two components.

Tungsten

Tungsten has the lowest expansion coefficient of the metals to be sealed. It requires sealing glasses with coefficients of thermal expansion (α) between 4.0 and 4.4×10^{-6}/K, such as borosilicate glasses (Glass No. 3 in Table 6.3). They are especially well-suited for the manufacture of high temperature incandescent and discharge lamps (flash tubes). If operating temperatures go as high as 750°C, alkaline earth-alumina-silicate glasses (Glass No. 4) have to be used. Their transformation temperatures can lie as much as 200°C above that of other multi-component glasses. Since they are alkali-free to a high degree and therefore highly insulating glasses, they are suitable for halogen–tungsten lamps and also as core substrate for high power metal film resistors.

Molybdenum

Molybdenum is a classic sealing material like tungsten. Owing to its high electrical conductivity, it has maintained its popularity over the Fe–Ni–Co alloys. There are a number of sealing glasses for molybdenum and Fe–Ni–Co alloys with α-values around 5×10^{-6}/K. The alkali-borosilicate glasses (Glass No. 5) are particularly suitable if special electrical qualities are not required. If high insulating values must be maintained up to 300°C (such as for internal glass parts of high power lamps), glasses with a higher content of B_2O_3 and a lower content of Na_2O are used.

Kovar

An even higher content of B_2O_3 (17–23%) is necessary in order to reduce the transformation temperature of the glasses below 510°C so they can be sealed to Fe–Ni–Co alloys (e.g., Kovar: 28% Ni, 18% Co, 54% Fe) in spite of the kink at 410°C in their expansion curves. This alloy is very important for glass-to-

Table 6.3 Examples of special glasses used in electrotechnology and lamp manufacturing

Glass number	1	2	3	4	5	6	7	8	9	10
Manufacturer Code number	Osram 452 (Fused silica)	Corning 7913 (Vycor)	Schott 8487	Schott 8252	Schott 8412	Schott 8250	Osram 125	Schott 8531	Osram 713	Corning 7251
Transformation temperature T_g (°C)	≈1100	≈1050	523	725	565	495	435	435	530	543
Temperatures (°C) at viscosity of: $10^{1.3}$	1150	1020	535	725	565	507	429	430	550	544
$10^{7.6}$ (dPas)	1650	1530	760	935	782	715	635	585	787	780
10^4	—	—	1135	1250	1165	1060	1000	818	1224	1167
Coefficient of thermal expansion $\alpha_{20-300°C}$ (10^{-6}/K)	0.54	0.75	4.0	4.6	4.9	5.0	9.8	9.1	4.4	3.7
Density (g/cm³)	2.21	2.18	2.25	2.63	2.34	2.28	2.86	4.38	2.27	2.26
Elasticity modulus, E (10^3 N/mm²)	66	68	64	81	73	64	—	52	59	65

Property										
Specific electrical resistance, δ (Ω cm)* log δ at 250°C	—	9.7	8.3	—	7.4	10.3	8.7	11.4	8.3	8.1
log δ at 350°C	—	8.1	6.9	12	6.0	8.5	7.1	9.8	6.9	6.6
Temperature at $\delta = 10^8$ Ω cm (°C)	510	358	275	640	215	384	280	448	278	—
Loss factor† at 1 MHz (10^4 tan δ)	—	4	36	11	80	22	—	9	—	45
Composition (wt %):										
SiO_2	99.9	96	75.1	60	74.9	68.7	61.4	34.3	72.9	78
Al_2O_3	0.005	<0.3	1.3	14.5	5.2	3.0	2.0	—	4.5	2
B_2O_3	—	<3.5	16.7	4/5	10.8	18.6	—	—	14.5	15
Na_2O	<0.001	<0.03	4.3	—	7.0	0.8	6.7	—	3.5	5
K_2O	<0.001	—	1.4	—	—	7.5	7.7	5.6	2.4	—
MgO	—	—	0.4	2.0	—	Li_2O:0.6	—	—	—	—
CaO	<0.001	—	0.7	10.0	1.4	—	—	—	—	—
BaO	—	—	—	9.0	1.0	—	—	—	1.2	—
ZnO	—	—	—	—	—	0.6	—	—	—	—
PbO	—	—	—	—	—	—	21.7	59.6	—	—

*The logarithm of δ is listed instead of δ (log δ = 8.3, instead of $\delta = 10^{8.3}$).
†The phase angle between current and voltage is 90° for an ideal dielectric material; in practice it is 90° − δ. The (very small) angle δ is the loss angle and the relationship of the actual to idle power is the loss factor tan δ.

metal seals due to its extremely low thermal expansion and its high degree of corrosion resistance. A number of glasses with special properties have been designed for use with this alloy. Some of these glasses also have special qualities such as low X-ray absorption, high ultraviolet transmission, and high electrical insulation (Glass No. 6).

Lead glasses

These fulfil a number of important functions in electrotechnology and electronics. Their favorable electrical qualities are based on the fact that under an electric field the large, almost immobile, lead ions block the migration of the alkali ions within the glass network, thus reducing conductivity and dielectric losses. Lead glasses (Glass No. 7) are important in the manufacture of stems and bases for incandescent and discharge lamps, for television tubes, and for many other types of vacuum tubes. They are compatible with numerous Ni–Fe–(Cr) alloys and copper clad wire.

In electronics, high lead content glasses are used chiefly for the encapsulation of diodes and other components, such as precision resistors and ceramic or tantalum capacitors (Fig. 6.7). A highly insulating, alkali-free lead glass (with about 60%

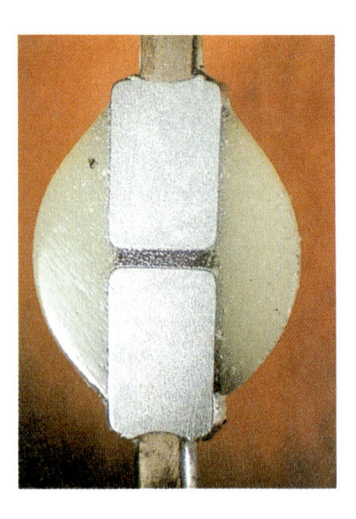

Fig. 6.7 Polished section of a glass-encapsulated silicon diode.

PbO) has been developed for alkali-sensitive silicone diodes. It has a very low sealing temperature (Glass No. 8). In order to create an hermetic seal, and hence provide chemical and mechanical protection for the components, precut sleeves of precision tubing are slid over the contact wire/semiconductor chip assembly and then sealed in a protective gas atmosphere using heating coils or conveyor furnaces.

Infrared absorbing glasses

In addition to flame and heating coils, infrared (IR) heat transfer (mostly from quartz–iodine lamps) is widely used for hot glass processing. Glasses containing several percent iron oxide (FeO) were developed for this purpose giving them a greenish tint. They absorb infrared and red light. Contrary to flame heating, hot shaping with infrared radiation can be conducted in a neutral or reducing atmosphere. This is advantageous for the hermetic encapsulation of electrical components in special gas atmospheres.

This process is used chiefly for the manufacture of reed switches (Fig. 6.8). These consist of two special magnetically activated alloy contact studs which are hermetically sealed to the ends of the glass envelope. These glasses are also used as sealing glasses for Ni–Fe alloys.

6.4.2 Glasses for television tubes

Television tubes require glasses with a variety of properties. They must be sealable to metal electrodes carrying high voltage (anode button) and to the conductors supplying the heating power and control voltages to the electron-beam gun. In addition, good X-ray absorption by the glass is of prime concern (Fig. 6.9). The glasses must have the required amounts of heavy oxides in their composition (BaO, PbO, SrO), in order to almost completely prevent X-rays produced in TV tubes from escaping. Glasses having different optical absorption characteristics must be used, depending on the wall thickness of the tube parts (panel, funnel (cone), neck). There are also special requirements for the glass used for the neck regarding high voltage resistance. Alkali–barium–silicate glasses are suitable for the picture screens (panels), and lead glasses are used

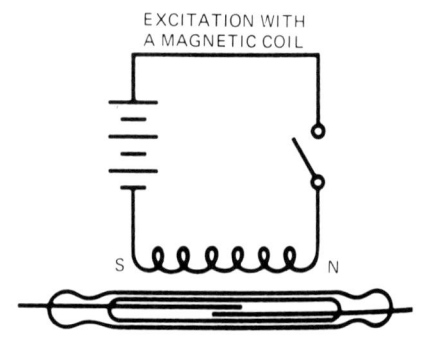

Fig. 6.8 Operation principle of a reed switch.

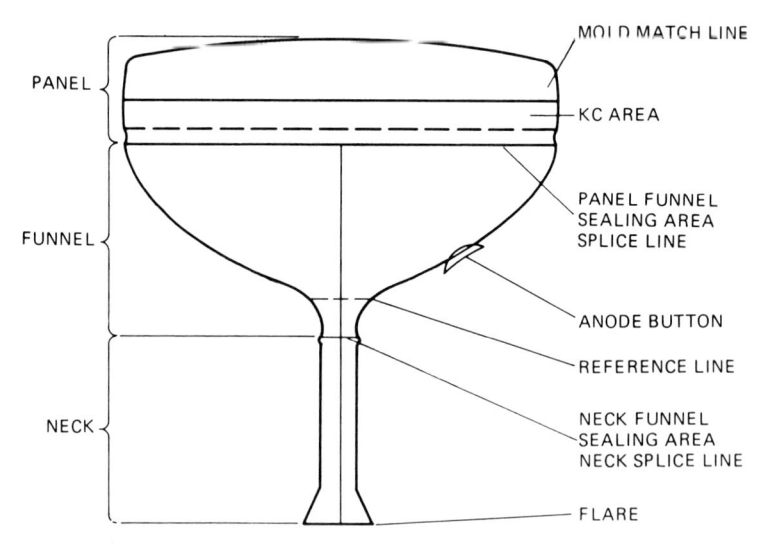

Fig. 6.9 Schematic design of a television tube.

for the funnel and neck, based on their optimum shaping properties and the best possible glass quality (freedom from knots, cord and stones).

6.4.3 Glasses for X-ray tubes, transmitting and image-intensifying tubes

For decades, glasses with a high content of barium- or zinc-oxide were used to make X-ray tubes. They resulted in relatively high absorption losses for X-rays. Only with the advent of newer glasses developed by Schott that contain elements with atomic numbers below 20 (atomic weights below 40) has satisfactory minimum X-ray absorption been achieved, while still meeting the requirements of the necessary glass-to-metal seals. These glasses are similar to Glass No. 6 in Table 6.3 and are compatible with molybdenum and Kovar alloys (e.g., *Vacon 10*). Since bubble and striae quality specifications are very stringent, they are melted in processes similar to those used for optical glass (see The manufacture of optical glass, Section 6.6.5).

The same glasses are also used for image intensifier and image converter tubes, vidicons, mercury switches and high-power transmitting tubes, which require high insulating capability and low dielectric loss.

6.4.4 Glasses for soldering and passivation

As with metallic solders, glass solders are glasses with a very low melting temperature that are used to join glass to other glasses, metals or ceramics with as little thermal impact as possible to the materials to be joined. When glass itself is a component to be joined, the glass solder must flow and wet at a temperature well below that at which the glass component would deform. Generally this temperature is determined by the transformation temperature of the glasses to be joined. Such solder glasses generally flow and wet at temperatures between 450 and 550°C. In order to do so, their viscosities at these soldering temperatures must lie between 10^4 P and 10^6 P (1 P = 1 dPa s = 10^{-1} N s/m^2).

As with sealing glasses, the thermal expansion of soldering glasses is determined by the components to be soldered. Since the solder is normally mechanically the weaker part, it is desirable to obtain a slightly compressive residual stress in the solder after the cooling cycle.

Glass solders usually come in powdered form (or sometimes in sintered preforms) with grain sizes of 60 μm or less. For an immediate application of the solder to the materials to be soldered, the powder is usually mixed with water or methanol to an easy-to-use suspension or paste. By using nitrocellulose dissolved in amyl acetate as a carrier, the powder will adhere, even after the solution dries. The nitrocellulose binder evaporates before the sintering phase begins.

Solutions may be applied by spraying, silk screening, or extrusion of a solder bead. The pre-pasted surfaces are assembled and the hot soldering process is carried out via an adequate temperature–time program.

Glass soldering is usually done when direct fusion is not technically feasible. Examples include picture tubes for colored televisions containing many components, and flat numerical displays (liquid crystal and gas discharge displays).

Glassy solders

These do not noticeably crystallize during the soldering process (which lasts from several minutes to an hour). In this case, the viscosity is reversible during heating and cooling (thermoplastic glass solders). For this reason, assemblies of parts joined by glassy solders can be corrected or separated when reheated.

Glass solders showing these characteristics are generally lead-borate glasses with 60–90% PbO by weight. To improve their chemical resistance, they usually also contain SiO_2 and Al_2O_3.

Crystallizing solders

These generally maintain their glassy character until they reach the soldering temperature. If the soldering temperature is sustained, they transform into a glass-crystalline or ceramic-like body due to the formation of crystals. This process results in an irreversible joint which prevents any corrective operation on the assembly even after reheating. Such fused joints can only be separated by chemical dissolution of the solder layer.

Crystallizing solders differ from glass solders primarily in their zinc oxide content (8–25% by weight). If soldering temperatures above 550°C are permitted (for ceramics or metals), zinc-borate and silicon-borate glasses (50–65% ZnO, 0–15% SiO_2, 20–35% B_2O_3 by weight) can also be used.

Composite glass solders

The development of glass solders with ever lower soldering temperatures is limited by the general rule (Section 2.3.1) that, at the same time, this leads to glasses with increasing coefficients of thermal expansion. While this effect is smaller in crystallizing solders, a far more effective method entails the addition of inert (non-reacting) fillers with low or negative α-values to the powdered glass solder. Included among such suitable fillers are the zircon mineral ($ZrSiO_4$) and β-eucryptite ($Li_2O \cdot Al_2O_3 \cdot 2SiO_2$), which also play a major role in the manufacture of glass-ceramics (compare with Glass-ceramics, Section 6.8). Such composite glass solders are preferably used as stable glass solders. The amount of filler which can be added to the batch is limited, however, by the unavoidable reduction of flow capability during the soldering process.

Passivation glasses

These glass solder-related zinc-silicoborate and lead-alumina-silicate glasses are used for chemical and mechanical protection of semiconductor surfaces, especially silicon components. They are applied by powder-paste technology (grain size approx. 10–20 μm) and melted directly onto the circuitry as a 10–50 μm thick layer or as an encapsulating cover over the circuitry. It is of utmost importance that these glasses are absolutely free of alkalis (as these can disturb the semiconductor function), and maintain a high dielectric and mechanical resistance. Alkali-free composite glasses have proven effective in matching the thermal expansion of silicon ($\alpha = 3.3 \times 10^{-6}$/K) at moderate melting temperatures (700–800°C).

Shielding from alkali migration is usually also required when semiconductors are bonded to glass substrates containing alkalis as, for instance, in the transparent electrode layers in liquid crystal displays. To accomplish this, an SiO_2 or other oxide layer is usually deposited by dip-coating or spray-coating or by precipitation from gaseous reactions.

6.4.5 Sintered glass parts

Small glass components that are regularly used in large quantities in electronics cannot be economically produced with sufficient precision by hot glass-forming processes. Such small components are usually made by sintering from either dry or wet-milled glasses. The milled glass powder with an average grain size of about 50 μm is mixed with plasticizing organic

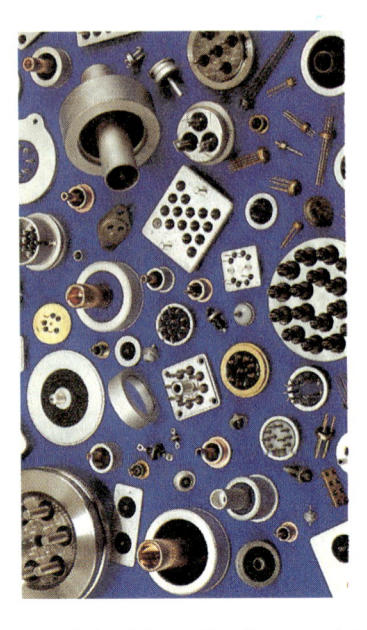

Fig. 6.10 Glass-to-metal feedthroughs for transistors, quartz-crystals and diodes.

binding agents to form a granulate. Then preforms are dry-pressed into their final shape and sintered at temperatures up to 600–700°C, whereby the glass preform solidifies into a mechanically stable gas-tight glass body. Dimensional tolerances can be held within 2%. Such granulates are made of many different types of special glasses, especially sealing glasses (Sealing glasses, Section 6.4.1).

A primary application of sintered glass parts is in insulated feedthroughs for electrical leads in hermetically sealed housings (Fig. 6.10). The sizes of these housings can vary from a few cubic millimeters for semiconductor components, up to several cubic meters for nuclear power facilities (Fig. 6.11). They usually consist of a metal ring surrounding which is sealed to a sintered glass component with one (or several) electrical conductor(s) sealed into it. Other sintered glass products include insulators, precision stand-off sleeves, bases and headers for transistors, relays, quartz crystal oscillators, etc.

Glass foils, only 0.2 to 0.8 mm thick, can be cut, stamped or drilled to size and then converted into sintered glass-ceramic.

Fig. 6.11 Glass-to-metal sealed reactor feedthroughs in the nuclear-powered ship, *Otto Hahn*.

These parts are also used in electronic and electrical applications.

6.4.6 Glasses for high-voltage insulators

These articles are in general pressed from soda-lime glass (sometimes with a certain PbO content) and are usually thermally toughened to increase their mechanical strength (Safety glass, Section 4.4.5). For voltages in excess of 20 000 V, borosilicate glass is preferred because of its resistance to thermal shock and moisture. The breakdown voltage of the glasses is approximately 450 000 V/cm at 50 Hz. They can be used successfully in competition with ceramic insulators in overhead railway lines and in overland high-voltage power lines. The insulators are suspended in such a fashion that the individual glass elements are not subject to any tensile stress.

6.4.7 Ultrasonic delay lines

The transmission of electrical signals often requires precisely defined delays in signal transfer. This is difficult to achieve by purely electrical means. The problem arises, for example, in the creation of color pictures in television tubes. The three electron beams for the red, green and blue color components scan the screen in lines, thereby determining the brightness and color composition of each point on the screen. The beams must be synchronized in sequence at 64 microsecond intervals (6.4×10^{-5} seconds), which is the time needed to scan one line. This delay is very long compared with the speed of these electric signals and is achieved by transforming the electric signal into an ultrasonic signal through a small glass plate with several parallel lateral surfaces. The electrical signal is then sent through an electromechanical transformer located at one lateral surface to generate the ultrasonic wave. After being reflected back and forth several times in the glass, it is then reconverted into an electrical signal at the exit surface. Special lead glasses can fulfil the requirements of low ultrasonic loss and minimum temperature dependence of the ultrasound wave propagation better than any other material. Glass delay lines are also used in computers and radar equipment. If temperature stability of the delay time is not very important, quartz glass (Fused silica, Section 6.1) can be used for its low attenuation of sound waves.

6.4.8 Electron conductive glasses

The previous sections and Table 6.3 demonstrate that the specific electrical resistance of glasses (ρ) increases with a decreasing alkali ion content. (ρ is the electrical resistance of a 1 cm cube of the glass.) At normal temperatures, values lie between 10^{12} Ω cm (very high alkali content glasses) and 10^{20} Ω cm (fused silica). The volume conductivity (as opposed to the surface conductivity which is caused by adsorbed water) is based on the ability of the ions to migrate in an electrical field. Migration is very low at room temperature, but increases quickly as temperature rises. Under sustained direct current, the mobile metallic ions can collect at the cathode and cause polarizing effects.

There are conductive, alkali-free glasses with a resistance of only 10^4 to 10^{10} Ω cm at normal temperatures. They conduct electrons because they contain polyvalent elements. This is the case with vanadium-phosphate glasses (V_2O_5–P_2O_5), where both V^{4+} and V^{5+} ions are present when there is an oxygen shortage. The electrical conductivity is a result of electrons jumping from one ion to another. Typical semiconductor effects are also present in chalcogenide glasses. The elements sulfur (S), selenium (Se) and tellurium (Te) belong to the chalcogens. In combination with arsenic, antimony, germanium and/or halides, low melting-temperature chalcogenide glasses can be made with the exclusion of oxygen. These are of great interest because of their infrared transmittance and also their sudden change from conditions of low conductivity to high conductivity at certain threshold voltage values. Development in this field has not yet reached technical maturity, however.

Semiconducting oxide glasses are of particular interest in the construction of photomultipliers (secondary electron multipliers). For example, if a voltage potential of about 500 volts per centimeter length is applied between the ends of a small diameter tube made of such a glass, a stream of electrons flows within the glass. Electrons set free from a photocathode by the impact of a light beam and directed into the tube release several secondary electrons when they hit the wall of the glass. These secondary electrons are accelerated by the electric field and each one of them produces the same effect. If this process is repeated along the channel only ten times, a million-fold more electrons arrive at the receiver end (e.g., fluorescent screen) than start at the entrance of this 'microchannel'.

This principle has been further developed for image intensification and transformation of weak or invisible (infrared, ultraviolet, X-ray) radiation into visible light. The required large number of image points is achieved by using an array of thin hollow glass fibers, the length of which is about ten times the internal width. In order to make such hollow glass fibers, sheathed fibers (Glass fiber optics, Section 6.7.3) having an acid-soluble glass core are drawn, fused into a bundle, and cut into disks. The core glass is then leached out with acid. If the glass itself is not a semiconductor, it can be coated with conductive oxides in a gas reaction process. The cladding glass, however, is usually a lead glass which forms a conductive surface layer when exposed to hot reducing gases, such as hydrogen.

6.4.9 Lamp glasses

The most important requirement of a glass used in electric bulbs is to form a transparent envelope around the actual light source (such as the tungsten filament in incandescent lamps) and to be impermeable to gases at high temperatures. The current is supplied through metallic conductors which are sealed into the glass envelope. In this configuration a vacuum-tight seal is best achieved when both the glass and the metal have similar coefficients of expansion (Fig. 6.12).

For low power lamps, copper clad wire is usually used to conduct the current. This wire has an iron–nickel alloy core sheathed with a copper layer of 20–30% of the cross section. This metal combination has a radial coefficient of expansion of 9.0 to 9.5×10^{-6}/K. Accordingly, soft glasses having α-values between 9 and 10×10^{-6}/K are required for the bulbs.

The most widely used type of incandescent lamp is the common pear shape. The bulb is made of a simple soda-alkaline earth-silicate glass and is machine blown on high throughput automatic machines (e.g., ribbon machines; The shaping of hollowware, Section 5.2). The electrical insulation of the bulb glass at operating temperatures is not high enough, however, so a highly insulating glass must be used for the 'socket' through which the sealed-in wires are fed. Lead glasses with a 20 to 30% PbO content are used for this application (Glass No. 7 in Table 6.3). These glasses are drawn in tube form using Danner or Vello machines (The drawing process, Section 5.3). To produce frosted bulbs, the inside surface of the glass is etched with a mixture of hydrofluoric acid and fluorides to obtain the desired degree of surface etch.

Various other incandescent bulbs are made using the same process (for example, bulbs for vehicles, photography and microscopy).

Fluorescent lamps are low pressure mercury-discharge lamps which have a fluorescent substance coating the inside of the tube. In the tube, made of soda-alkaline earth-silicate glass, mercury vapor is excited to produce ultraviolet radiation, which is transformed into visible light when it strikes the fluorescent material coated on the inside of the glass tube. Due to the high ultraviolet absorption of the fluorescent material as well as the glass, no ultraviolet light is emitted from the tube. Here again, lead glass is used to seal the electric leads into the sockets of the tube.

Soft glass does not have sufficient thermal shock resistance to be used in high power lamps. Hard glasses with coefficients of thermal expansion below 5×10^{-6}/K are used instead (e.g.,

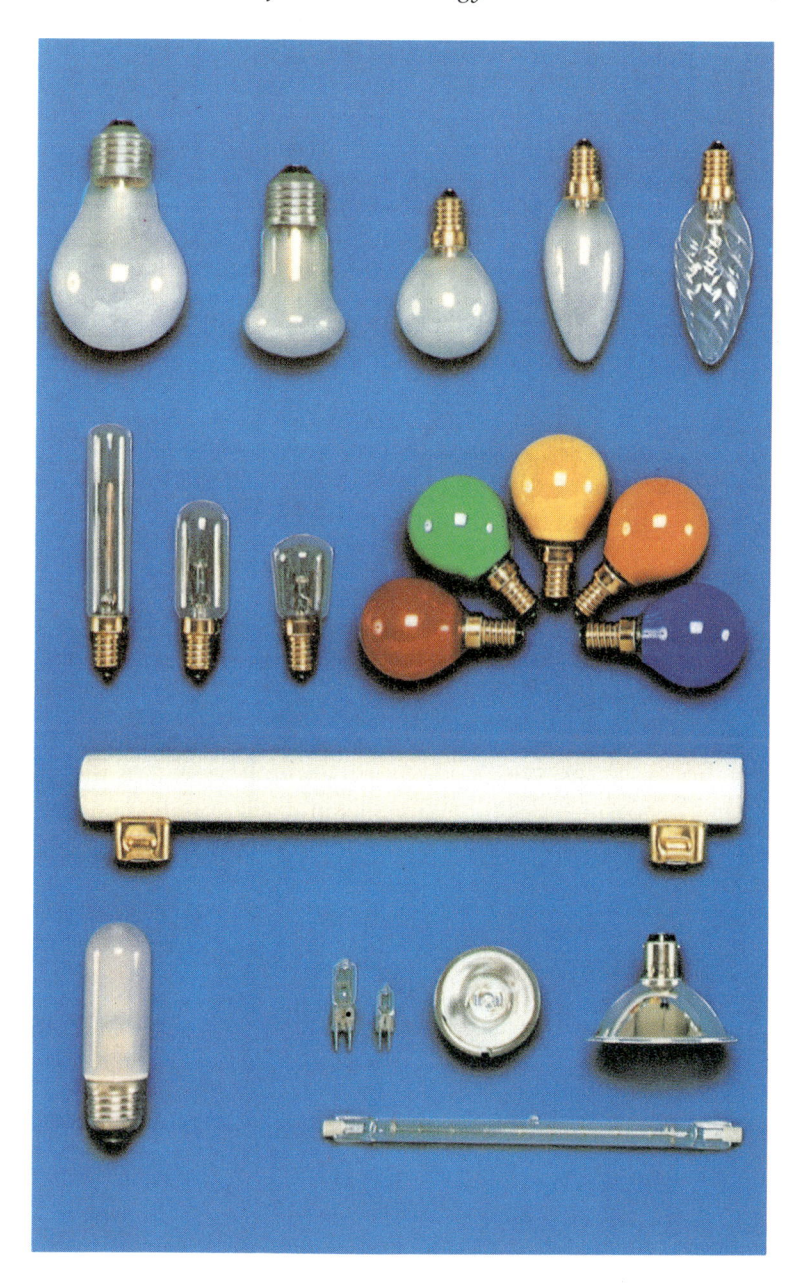

Glasses for electrotechnology and electronics

Fig. 6.12 Variety of bulbs and lamps made by Osram.

Glass No. 9 in Table 6.3). Molybdenum, tungsten, and Fe–Ni–Co alloys are the conductor materials used for glasses within this expansion range.

A number of highly focusing incandescent lamps, such as the CONCENTRA® or sealed beam headlights used in the USA, are made of pressed glass parts (Glass No. 10 in Table 6.3). Both the reflector component and the cover are made of borosilicate glass with 12–15% B_2O_3. The reflector is mirror-coated in the lamp factory (e.g., aluminum is vacuum-deposited on the glass reflector), the tungsten filament assembly is inserted, and the cover is fused to the reflector rim.

In projection bulbs, the bulb temperature exceeds both the annealing temperature and the softening point of soft glasses and most borosilicate glasses. For this application bulbs must be made of glasses with transformation temperatures above 700°C. The glasses used are alkaline earth-alumina-silicate (similar to Glass No. 4 in Table 6.3) with 14–24% Al_2O_3 and 4–9% $B_2O_3 + P_2O_5$. The base is made of highly insulating glass, usually tungsten sealing glass (similar to Glass No. 3 in Table 6.3).

Halogen incandescent lamps have halogens added to their filler gas. The halogens inhibit the blackening of the surface of the glass bulb usually caused by vaporized tungsten. Owing to the higher bulb temperature required for the so-called halogen cycle process, these lamps must be much smaller than normal incandescent bulbs of similar power. For this reason, quartz glass or Vycor (Fused silica, Section 6.1) are used. Owing to the vast differences between the thermal expansions of quartz glass/Vycor and molybdenum (5.4×10^{-6}/K) a vacuum-tight contact wire seal is not possible. The electric feedthroughs are therefore made of a 20–40 μm thick molybdenum foil pinched between the softened glass when the socket is formed.

Alkaline earth-alumina-silicate glasses (similar to Glass No. 4 in Table 6.3) are also used for low power halogen incandescent lamps. Their transformation temperatures lie between 700 and 800°C, and they are fused directly to the molybdenum wires. With few exceptions, the glass tubes are machine drawn.

In high pressure mercury discharge lamps, the radiation is generated in a tube of highly ultraviolet transmitting quartz glass. Here too, a molybdenum foil is used as the electrical conductor. The discharge tube is hermetically sealed into an

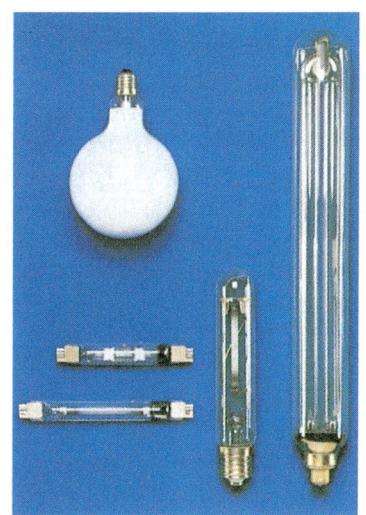

Glasses for electrotechnology and electronics 147

Fig. 6.13 Discharge lamps.

enveloping bulb, which is internally coated with fluorescent material for converting the ultraviolet radiation into visible light (Fig. 6.13).

Low power lamps have their envelope bulbs made of soda-alkaline earth glass, and sockets made of lead glass. Power lamps above 250 watts require bulbs made of borosilicate glass and sockets made of highly-insulating borosilicate glass.

The discharge tubes of special extra-high pressure discharge lamps (e.g., xenon lamps) become very hot (between 1000 and 1200°C); therefore, quartz glass is used. Lamps with relatively low current loads are manufactured with molybdenum foils as electric conductors, while those with high current loads require tungsten rods. In order to overcome the high difference in thermal expansion between quartz glass and tungsten, two or three transition sealing glasses with coefficients of expansion of 1.3, 1.8 and 2.3×10^{-6}/K are used to make a 'graded seal'. The glass with the highest expansion is fused to tungsten.

In sodium discharge lamps, sodium vapor is excited to the discharge point. The low pressure lamp emits a nearly pure monochromatic yellow light. At high temperatures, sodium vapor destroys standard technical glasses. For this reason, special glasses have been developed, the most widely used of which are the barium-borate glasses with a low silica content. Most of these glasses are either difficult to work (short), or are insufficiently resistant to weathering, hence a two-layer (flashed) glass is used for the bulb. The soft glass tube (alkali-alkaline earth-silicate glass) is internally coated with a thin layer (50 to 100 μm) of sodium-resistant glass. This flashed glass tubing is mechanically drawn.

The high pressure sodium lamp produces a yellowish-white light, and the discharge tube heats up to about 1100°C. Temperature and chemical attack exclude the use of glass altogether and instead, sintered alumina (Al_2O_3) is used. The exterior bulb, however, is made of borosilicate glass.

Spectral lamps are used to produce line spectra and consist of a discharge chamber with a base glass and the metal to be excited, as well as an exterior bulb made of soft glass. Quartz glass or special glass is used for the discharge chamber, depending on the type of discharge and metal vapor filling.

High transmission at the mercury line (254 nm) is required for ultraviolet radiators which are used in sterilizers and other medical equipment. Quartz glass is therefore used. On the other hand, radiation below 280 nm is not desired in ultraviolet radiation equipment used for cosmetic-therapeutic purposes. Soda-alkali earth glasses are used for such applications. Their ultraviolet transmission can be controlled and reduced, e.g., by adding Fe_2O_3 to the composition. The shortwave ultraviolet radiation can also be suppressed by a thin oxide coating on the bulb.

Soft glass is used in high voltage fluorescent tubes containing inert gases such as neon. These are used in the familiar colorful neon signs seen in advertising. The lamps contain small quantities of mercury to excite the gas discharge. Since ultraviolet radiation and mercury attack will discolor normal glasses at the high operating voltages (up to 6 kV), these glasses must be specifically adjusted to their application.

Either soft glasses or borosilicate glasses are used to produce flash bulbs. The glass used depends upon the temperature and pressure resulting from the flash.

Colored lamps can be made in different ways. Glasses colored in the bulk use CdS for yellow, Co and Cu for blue, Cr for green and Se for red. There are also surface-tinted glasses.

They are stained with silver or copper compositions to obtain yellow, red and brown (Painting, Section 5.8.3). The inner surfaces can also be coated with colored powders, or colored lacquer can be applied to the outer surfaces.

Infrared lamps are incandescent lamps working at low filament temperatures and therefore emitting a relatively high amount of infrared radiation. The bulbs are usually made of borosilicate glass with sealed in molybdenum or tungsten wires.

6.5 ELECTRODE GLASSES

The knowledge of the pH-value of liquids plays a major role in the control of various chemico-technical processes, such as monitoring drinking water, but also in medical labs, dairies and many other areas. pH-values are a measure of the hydrogen ion concentration in solutions, which in turn indicates their acidity and alkalinity.

Today, pH-values are almost exclusively measured by glass electrode chains. A potential difference (Galvani voltage) is formed at the interface between a glass membrane (approximately 0.5 mm thick) and a liquid electrolyte (ion-conductor). The potential difference between the glass and the aqueous solution is dependent upon the pH value and, for certain glass compositions, the alkali-ion concentration of the solution (sodium ion concentration, in particular). This effect is utilized to measure the pH, pNa, etc. of a solution.

As glass in itself cannot serve as an electrode, the two-phase glass/solution system is expanded into a multiphase system by adding a metallic end phase (electron conductors). Figure 6.14 is a schematic drawing of such a glass electrode measuring chain. Silver/silver chloride, calomel, and Thalamid® electrodes (Schott) are used as internal conductor and reference electrode. The Thalamid electrodes have the advantage of good reproducibility and high operating temperatures (up to 135°C).

6.6 OPTICAL AND OPHTHALMIC GLASS

6.6.1 Properties and classification of optical glasses

The most important properties of glasses for optical applications are their refraction of light and their dispersion of color.

150   *Special glasses and their uses*

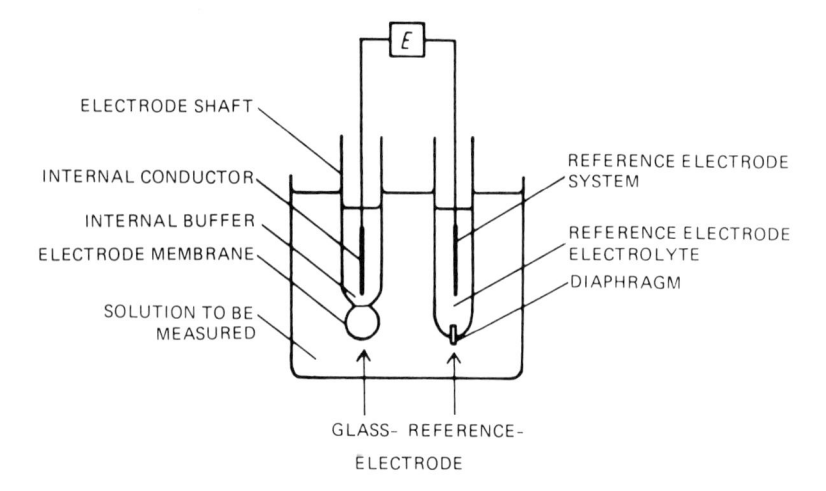

Fig. 6.14 Schematic diagram of a glass electrode measuring chain. The measured chain voltage, E, is composed of the voltages at the various phase boundaries (i.e. internal conductor/internal buffer, internal buffer/electrode membrane, etc.) which are all kept invariable except for the voltage of interest at the 'electrode membrane/solution-to-be-measured' interface. This makes E a function of the pH-value of the solution. The diaphragm prevents the reference electrode electrolyte from mixing with the solution to be measured, while ensuring that there is electrical contact.

Both are functions of refractive index (n_λ). The subscript, λ, indicates the dependency of the value n on the wavelength λ of the light. The refractive index of optical glasses increases from the red end of the spectrum towards the blue (i.e. the glasses have a normal dispersion when the refractive index increases with decreasing wavelength).

The higher the refractive index (n) the greater the bending of a ray of light striking the glass surface at an oblique angle. Reference points on the curve of n_λ vs. λ are generally identified by the wavelengths of several standard spectral lines of chemical elements (Fig. 6.15). The most frequently used spectral lines are the yellow helium line d at 587.6 nm, the green mercury line e at 546.1 nm, and the blue and red hydrogen lines F and C at 486.1 and 656.3 nm, respectively. The difference $n_F - n_C$ is called the main dispersion. The ratio $(n_d - 1)/(n_F - n_C)$ is called the Abbe number, ν_d. For optical

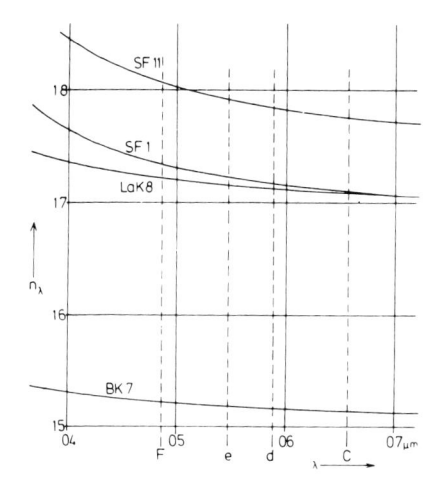

Fig. 6.15 Refractive index versus wavelength (dispersion) for various optical glasses.

classification, the optical glasses are entered on a chart, the abscissa of which is the Abbe number v_d, and the ordinate the refractive index, n_d (Fig. 6.16). The position of a glass in this diagram is usually designated as its 'optical position'.

For two different glasses with normal dispersion behavior, all their dispersion values coincide, if their n_d and v_d values are the same. For this reason, the entire refraction behavior of 'normal glasses' is defined by these two values. A low Abbe number, v_d, identifies a glass with high dispersion, with respect to its refractive index. Such glasses are called flint glasses.

Crown glasses, on the other hand, have relatively high Abbe numbers, with respect to their refractive indices. The Abbe numbers of the optical glasses lie between 20 and 90; the borderline between crown and flint glass by definition lies at an Abbe number of 50.

The historical development of the optical glasses was characterized by three successive eras. Prior to 1880, optical glasses consisted of simple crown glasses (soda-lime-silicate glasses) with relatively low dispersion, and simple flint glasses (lead-alkali-silicate glasses) with relatively high dispersions. Otto Schott conducted extensive glass trial melts using BaO, B_2O_3, and KHF_2 between 1880 and 1895, which resulted in a

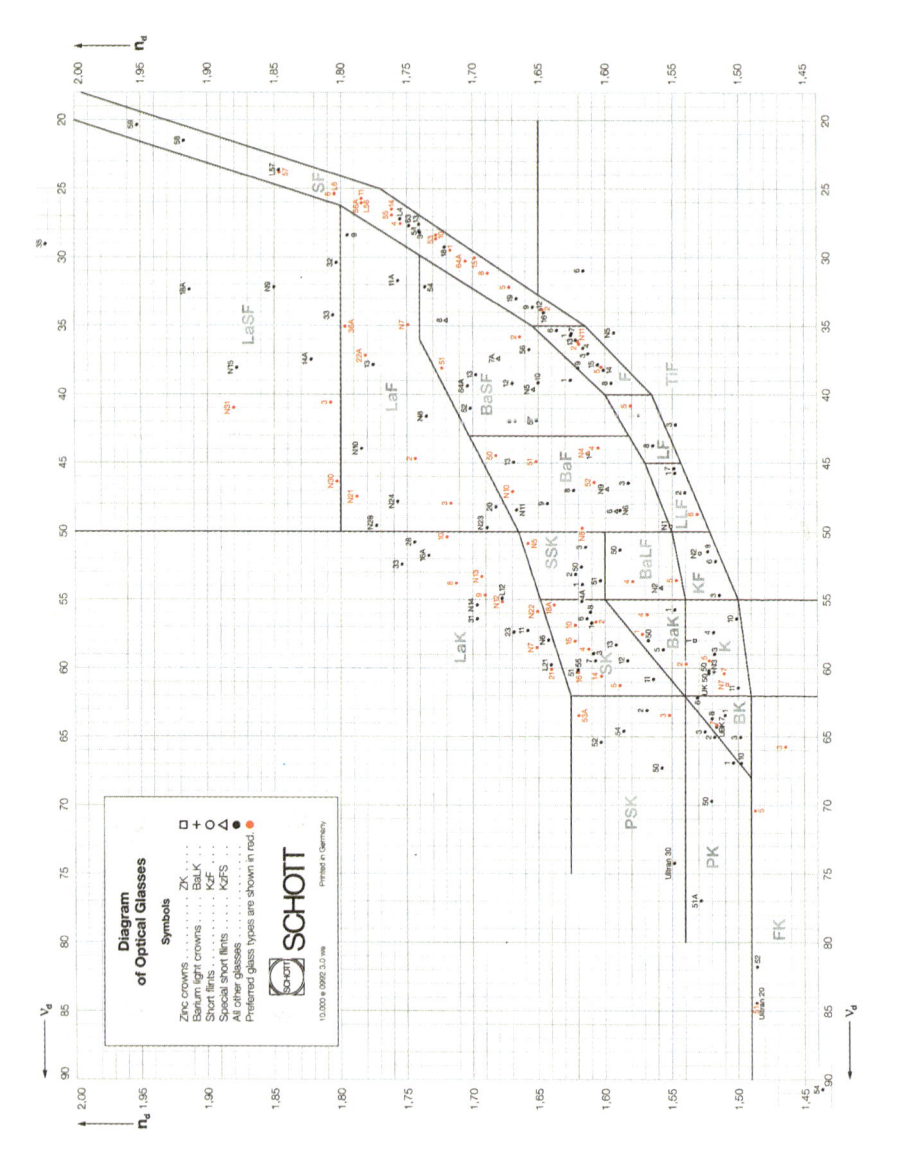

Fig. 6.16 Diagram of optical glasses. (B boron, Ba barium, F flint, K crown, L light, La lanthanum, LL double light, P phosphate, S heavy, SS double heavy, Ti deep)

The symbols indicate the location of the various optical glasses of each glass type in the n_d/ν_d diagram.

variety of new optical glasses. For example, barium oxide yields an unusually low dispersion at a relatively high refractive index. Boron oxide yields a low refractive index and very low dispersion, while the use of fluorine instead of oxygen also lowers the refractive index and dispersion.

In 1930 another era of new glass developments began. This work resulted in optical glasses containing rare earths, especially lanthanum. These glass types expanded the range of optical glasses into the area of high refractive index and high Abbe values, and were decisive in further eliminating image defects in lens systems.

The optical glasses are traditionally identified by combinations of the terms crown and flint with the terms heavy and light which denote high and low refractive indices. Supplementary designations are given to subgroups according to the chemical components which are important in determining their optical positions (for example, barium heavy flint (BaSF) or phosphorous heavy crown (PSK), see Fig. 6.16).

Barium crown glasses contain a large proportion of boron oxide and barium oxide, while their SiO_2 content is relatively low. Small additions of substances, such as aluminum oxide stabilize the glass against devitrification and weathering. Additions of aluminum oxide also inhibit fluoride crystallization in fluorine crown glasses.

The borosilicate crown glasses are equivalent to the technical borosilicate glasses. The calcium oxide of normal soda-lime silicate glass is replaced by boron oxide. The light and heavy flint glasses and barium crown glasses are characterized by low and high lead- and barium-contents respectively. Barium flint glasses contain both barium oxide and lead oxide; crown flint glasses contain calcium oxide and lead oxide, resulting in average dispersions.

The density of the glass decreases with increasing temperature, while both the dispersion curve of the glass and its ultraviolet absorption cut-off line shift (Fig. 6.15). Both of these affect the refractive index, so that it is possible to have glasses with either negative or positive temperature coefficients of n_λ. In athermalized glasses the optical light path (n_λ times glass path) is almost independent of the temperature. This is important in systems having large focal lengths, such as telescopes.

In the manufacture of high quality multi-lens systems, optical glasses are required not only for achromatic lenses, but also for apochromatic lens combinations. In an achromatic combination, the chromatic aberration is eliminated for two colors (such as red and blue); however, a residual error (secondary spectrum) remains in the intermediate range. In an apochromatic lens, this residual error is almost completely corrected as well. This is done by using special glasses with different partial dispersions. Although these glasses have the same n_d and ν_d values as the corresponding flint glasses, the difference of their refractive index between the blue and the ultraviolet is 'shorter' than in the corresponding flint glasses (hence, the name short flint).

Glass types with greatly differing partial dispersions are of particular interest in large diameter lens systems with long focal lengths. The resolution of such a system is greatly increased to the extent that even the slightest chromatic defect becomes noticeable.

The quality requirements of optical glasses today far exceed those of most other glasses. Among the most stringent requirements are freedom from striae, optical homogeneity (i.e. consistency of refractive index within a melt), the smallest possible bubble content, minimal absorption in defined spectral areas, and low birefringence caused by internal stresses in the glass. With today's technology, for example, it is possible to produce highly homogeneous glasses with a refractive index variance $\Delta n_d \leq 1 \times 10^{-6}$.

6.6.2 Transmission of radiation; color filters

Transmission, absorption and reflection are qualities which are equally important in optical glass, colored glass and filters, and colored ophthalmic glasses. When light strikes a sheet of glass it is partially reflected, partially absorbed, and partially transmitted (Fig. 4.11).

The degree of reflection, r, is the ratio of the intensities of light reflected from a surface to that of the incident light. The spectral transmittance, $\tau(\lambda)$, is the ratio of the intensity of light transmitted through the glass to that of the incident light ($\leq 5 \times 10^{-7}$).

The reflection loss of simple soda-lime-silicate glasses at perpendicular light incidence is approximately 4% at each surface; however, it increases greatly when the angle of incidence increases above 40° to almost 100% at a glancing incidence (90°). Absorption occurs when the quantum energy of the light is sufficient to excite the electrons bonded within the glass structure. In quartz glass and soda-lime glass, the bonding of oxygen atoms to silicon or to cations is so strong that electron excitement can only be caused by ultraviolet radiation.

Consequently, these glasses are completely transparent to the visible light spectrum. On the other hand, small concentrations of transition elements (see below) not only cause strong absorption bands in the ultraviolet region, but also broad bands in the visible and in the infrared spectral ranges. There are also several narrower absorption regions which can be attributed to hydroxyl (OH) groups and water molecules within the glass. All silicate glasses are opaque to infrared waves above 5 μm as a result of the natural oscillation of the Si–O groups which absorb these waves. Glasses for infrared optics must not contain SiO_2 and they must not contain any oxides at all for wavelengths over 6 μm. Instead, compounds of such elements as sulfur, selenium, arsenic, antimony and germanium are used. More recently, oxide-free fluoride glasses which are based on ZrF_4 and HfF_4, often with the addition of BaF_2 or ThF_4, are used in infrared optical systems. They are transparent between 0.2 and 9 μm and must be manufactured in a completely oxygen-free environment (see also Coloring agents, Section 2.4.9).

Coloration in glass, caused by absorption bands of transition elements (such as Cu, Ti, V, Cr, Mn, Fe, Co, Ni, etc.), varies greatly (Raw materials, Chapter 2). The structure of the host glass also influences these absorption bands to a certain extent. Each change in the bonding relationships in the glass structure affects the electron bonding and therefore the color characteristics (Fig. 6.17).

The addition of rare earths, on the other hand, creates narrow absorption bands. Glasses of this nature (mostly neodymium-containing) are used as laser sources (Fig. 6.18). Lasers (an acronym for 'Light Amplification by Stimulated Emission of Radiation') are light sources in which the stimulating energy of a pulsed light (xenon or krypton flash lamp) is transformed into monochromatic, coherent light of high intensity. They are used in applications such as machining of

156   *Special glasses and their uses*

Fig. 6.17 Color filter glass.

Fig. 6.18 Laser glass.

materials, measuring technology, optical data transmission and eye surgery. The stimulated neodymium ions in the laser glass emit infrared radiation at the 1.06 μm wavelength.

Another major group of color glasses is the one used for steep gradient color filters. Their production and their properties are fundamentally different from those of ion colored glasses. They show very little coloration in their molten cast form. The intense yellow, orange or red colors develop in a subsequent heat treatment depending on the doping level with chromophores and the duration of the heat treatment. Unlike ion colored glasses, these glasses absorb the incident light from the low ultraviolet to the absorption edge very strongly, while they transmit light of longer wavelengths with no loss.

Narrow band filters which transmit only a narrow spectral range (with half value widths < 0.5 μm) cannot easily be made with absorbing glasses nor can filters with a sharp transmission cutoff at longer wavelengths. Such requirements, however, can be met by interference filters which usually have many layers of oxides with different refractive indices. Some types also have several metallic coatings and are often used in combination with a color filter glass.

6.6.3 Ophthalmic glass (spectacle glass)

A number of glasses are used to correct the vision and to protect the eyes against undesirable light radiation. They are usually melted with great homogeneity in electrically heated furnaces.

Colorless ophthalmic crown glass (white spectacle crown) has a refractive index of 1.5230. The curves of the lens surfaces and the refractive index of the glass determine the optical power of the ophthalmic lens. The refractive power, D, is the reciprocal of the focal length (in meters) and is expressed in diopters (dpt.). Thus, an ophthalmic glass with a power of ± 2 dpt. has a focal length of 0.5 meters.

Between 20°C and 300°C ophthalmic crown glasses have a linear coefficient of thermal expansion of 9.5×10^{-6}/K. The viscosity/temperature dependence is comparable to that of soda-lime glass. Both factors (thermal expansion and viscosity behavior) are important for determining the fusibility of this ophthalmic crown glass to special flint glasses for multi-focal spectacle lenses (bifocal and trifocal glass) (Fig. 6.19).

As the power of the prescription increases (above ± 4 dpt.), the edge or the center thickness of crown glass spectacles become very thick and their weight makes them become unpleasant to wear. This led to the development of lighter, high-index glasses for thinner lenses with the same refractive power.

Crown glasses with a higher refractive index (Hi-Crown) are suitable for thin lenses in the lower and mid-diopter range. SF 64 has ushered in a series of high-index glasses that have proven extremely successful. Their properties have been furth-

INTERNAL FUSION   EXTERNAL FUSION

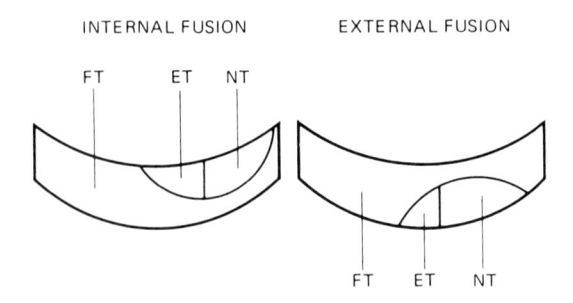

FT   ET   NT

FT   ET   NT

Fig. 6.19 Cross section of fused multifocal glasses. Internal fusion (left) and external fusion (right). In both processes the reading segment NT with a high refractive index (higher than the far sight segment FT) is first fused to the extension segment ET. In a second step, the two fused segments (ET and NT) are then fused to the far sight segment (FT) on either the concave or convex surface.

er improved with type BaSF 64, and LaSF 36A from which aesthetically attractive glasses, even for extremely high diopters, can be produced (Table 6.4).

For safety reasons, the mechanical strength of ophthalmic lenses has been improved by thermal and chemical strengthening processes to the point where some special glasses have eight times the strength of normal ophthalmic lenses.

The tinted absorption glasses are categorized into lightly tinted comfort glasses and sunglasses. Both types are ophthalmic crown glasses with a refractive index of 1.5230 and differ mainly in coloring components. The tinted absorption glasses (mainly with a tinge of pink), have transmission values between 85% and 60% (based on 2 mm glass thickness). Most

Table 6.4 Refractive indexes, n_e, and Abbe coefficients, ν_e, of high-index glasses

	Type of glass				
	Hi-Crown 42	Hi-Crown 45	SF64	BaSF 64	LaSF 36A
n_e	1.6040	1.6040	1.7064	1.7052	1.8000
ν_e	41.8	45.1	30.5	39.3	35.4

sunglasses have brown or gray or even green tones with transmission values between 65% and 20% (based on 2 mm glass thickness). The coloring oxides used are those of iron, cobalt, nickel, copper, manganese (Raw materials, Section 2.4), as well as combinations of iron oxide and selenium oxide. Fashionable tints, such as blue or yellow, complete the scale. Good sunglasses also have the lowest possible transmission of infrared and ultraviolet rays, which are not perceived by the eye, but in some cases are harmful.

Safety glasses, which protect against infrared radiation, are made with highly concentrated coloring oxides (Schott KG-glasses or ATHERMAL welding glasses made by Deutsche Spezialglas AG). These glasses have a well-defined reduced transmission, but are still sufficient for their purpose within the visible spectrum and also are virtually opaque in the spectral regions that are harmful to the eye.

When photochromic ophthalmic lenses are exposed to ultraviolet or shortwave infrared radiation, the transmission of visible light is automatically reduced. When the exposure ends, they return to their initial state within a short period (Fig. 6.20). This property, also called phototropy, is based on the formation of glassy or crystalline silver halide containing particles of submicroscopic size (diameter approximately 5–30 nm, concentration approximately 1:2000). In order to create this property in the glass, silver salts and halides (metal compounds with fluorine, chlorine, or bromine) are added to the batch (usually a borosilicate-base glass). Closely controlled thermal treatment during and after the melting process leads to the formation of the silver halide containing particles which are the basis for the phototropic behavior of the glass (Fig. 6.21).

The ophthalmic sector is one of the few areas where glass encounters competition from plastics. For simple, fashionably tinted anti-glare spectacles and industrial safety eyewear, materials such as acrylic glass (polymethylmethacrylate) or polycarbonates are used. These are not suitable for corrective eyewear, however. The major ophthalmic grade plastic material suitable for corrective lenses is CR-39 (abbreviation for Columbia Resin 39 made by the USA firm PPG; chemical designation: poly-diethylene-glycoldiallyl-bicarbonate). It is limited to weak prescriptions, however, as its low refractive index requires undesirably thick lenses in order to obtain the

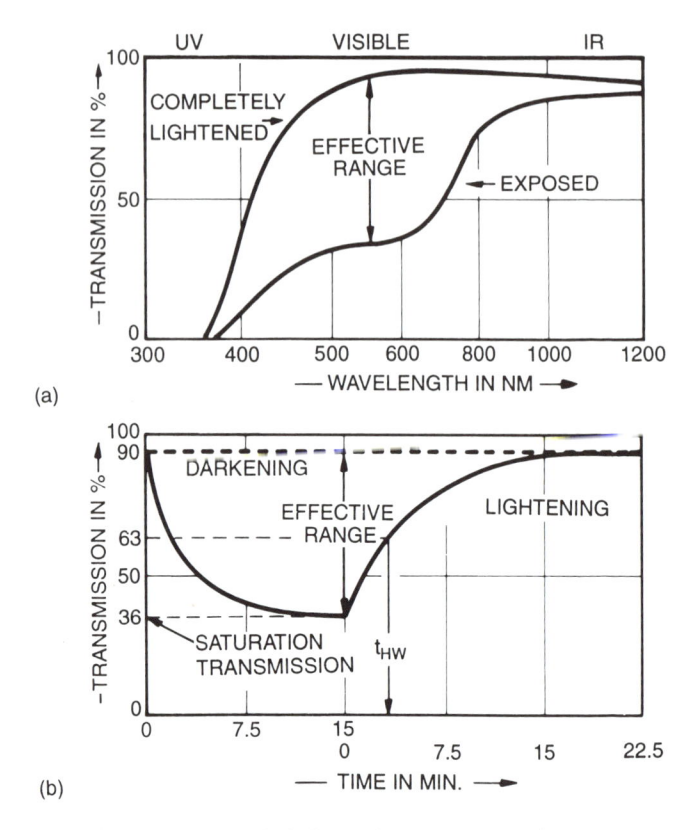

Fig. 6.20 Photochromic ophthalmic glass: (a) spectral transmission in exposed and unexposed condition; (b) darkening and recovery time.

Fig. 6.21 Photochromic ophthalmic glass with a gradient tint: unexposed (left); exposed (right).

necessary lens curvature in high corrections. In addition to good optical quality this material has adequate wipe and scratch resistance. In order to make the latter similar to that of glass, plastic safety spectacles (consisting partly of CR-39) are coated with thin layers of harder material. These coatings are usually precipitated from solutions of organic silicon compounds (methyl-polysiloxanes are often used). In some cases, even vacuum deposited glass has been used (Glasses with altered radiation characteristics, Section 4.4.1).

6.6.4 Special optical glasses for nuclear technology and radiation research

Radiation shielding glasses and radiation-resistant optical glasses form a separate group of glasses that are specifically developed for nuclear technology (Fig. 6.22). In shielding windows for 'hot cells', the high absorption of radioactive radiation by lead is exploited (Fig. 6.23). As these lead-containing glasses tend to discolor under α-and β-radiation, they are usually stabilized with cerium oxide. Schott's popular RS 520 G 5 radiation shielding glass, for example, contains 0.5% CeO_2. A sheet of this glass of a given thickness has about the same absorption as a sheet of solid lead of half that thickness.

Optical lead glasses are also used in the field of radiation research to detect and determine the energy of high-speed subatomic particles (electrons, positrons, cosmic rays, etc.). If such a particle travels through a medium (refractive index n) at a velocity of v which is faster than c/n, the velocity of light in that medium ($c =$ the velocity of light in a vacuum, $= 3 \times 10^5$ km/s), it produces electromagnetic radiation that expands as a conical wave, the so-called Cerenkov radiation (Fig. 6.24). This becomes visible as blue-violet light and can be monitored by using photomultipliers and impulse counters. Special glasses with refractive indices ≥ 1.7 and high transmission in the ultraviolet range meet the requirements for such observations for particle velocities $\geq 1.8 \times 10^5$ km/s. Certain atomic nuclei are capable of transforming photons from gamma radiation into electron–positron pairs, an effect that can also be detected and recorded by the Cerenkov effect. Similar verification methods are developed using scintillation glasses.

162   *Special glasses and their uses*

Fig. 6.22 Radiation shielding glass.

Fig. 6.23 'Hot cell' with radiation shielding window.

Through interaction with energy-rich radiation (particles or γ-quanta), certain atoms in glass are stimulated to emit light emission in the form of minute flashes (scintillations), which can be recorded and evaluated using highly sensitive photomultipliers.

Dosimeter glasses are used to measure the dosage of γ-rays and especially to monitor small doses of radiation to which

Fig. 6.24 Optical glass blocks for Cerenkov counters.

working staff may be exposed. There are also special glasses for neutron dosimetry (threshold value detectors) using cobalt- or silver-containing phosphate glasses, which form a color spot when exposed to radiation. Cobalt-containing glasses discolor under radiation between 425 and 370 nm. Silver phosphate glasses produce silver particles when exposed to radiation. The intensity of their fluorescence under stimulation with ultraviolet light is a measure for the number of produced structure defects and hence also of the radiation dose.

6.6.5 The manufacture of optical glass

The older process for melting optical glasses utilized the ceramic pot (or crucible) which was heated either alone or in batches of up to ten pots in a gas-heated, regenerative pot furnace. Glassmelts in large pots (up to 800 liters and above) were homogenized with special stirring equipment after the raw materiel batch had melted. After the desired homogeneity was obtained, the contents of the pot was cast into a rectangular steel mold and the glass would solidify as a block prior to annealing (Fig. 6.25).

Owing to the low yield of such a process and the tendency of many optical glasses to devitrify, pot melting is now scarcely done. Instead, larger quantities of optical glass are melted in continuous tank furnaces. Smaller quantities are melted in platinum or quartz crucibles (depending on the glass composition) in a method similar to the older pot process. Tank melting

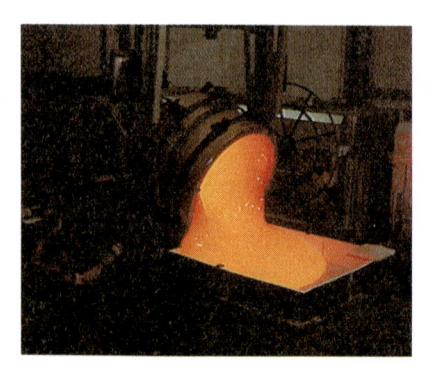

Fig. 6.25 Pouring from a pot melt.

of optical glasses is essentially the same as tank melting of other glass types (see Chapter 3).

Carefully controlled annealing is extremely important for optical glasses. In many cases fine corrections to the refractive index can be made during the annealing process.

As in pot melting, glass from a tank melt can also be used to produce blocks which can easily be tested for their optical qualities. In addition, glass rods with rectangular, circular, or triangular cross sections can be drawn in a continuous process (Fig. 6.26). These can be converted efficiently into optical glass components by the optical processing industry. Rotary table presses in conjunction with suitable glass-feed mechanisms (e.g., automatic shears) are used to manufacture blanks for prisms, lenses and spectacle lenses from molten glass taken continuously and directly out of the melting tank (Fig. 6.27).

6.6.6 Microspheres

Uses and properties

Clear and colored glass spheres with various refractive indices are used not only for fashion jewelry, but also for many technical applications.

Owing to their optical qualities, microspheres with diameters ≤ 0.2 mm are widely used in reflective signs and for projection screens. For this application microspheres with diameters between 0.10 mm and 0.16 mm and a refractive index of about

Optical and ophthalmic glass 165

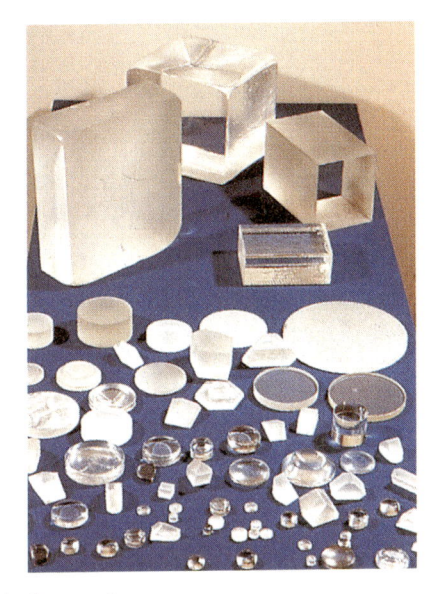

Fig. 6.26 Optical raw glass.

Fig. 6.27 Automated lens pressing.

1.70 are embedded in colored or fluorescent plastics or lacquer in order to enhance their reflective properties. Reflective foils with embedded glass spheres are also used for traffic signs and road markings especially for better sign recognition at night. The spheres act as optical lenses focusing incident light coming from a distance into a point close to the back of the spheres.

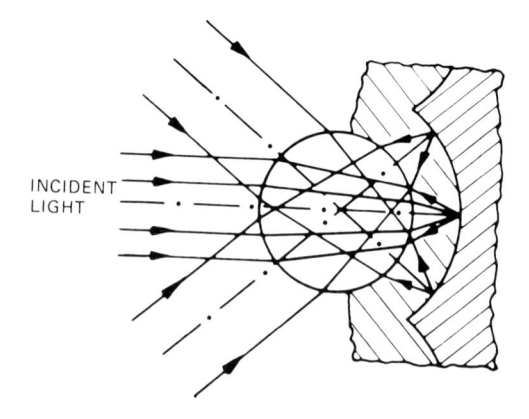

Fig. 6.28 Light path in a retroreflective system with glass spheres.

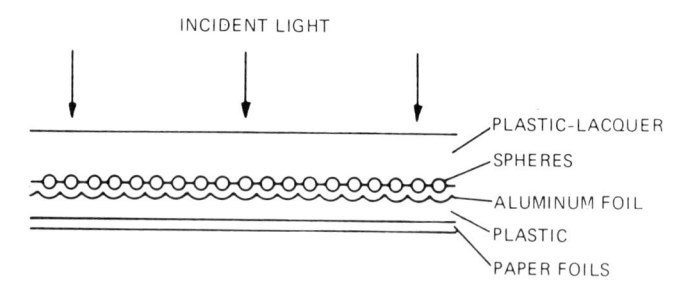

Fig. 6.29 Schematic design of a retroreflective foil.

When many spheres of approximately the same diameter are placed next to one another, all incident light is focused onto a single plane behind the spheres where a highly reflective metallic film is located. The incident light is thereby reflected back towards the source ('retroreflection') (Figs 6.28 and 6.29).

The distance of the focal point from the rear side of the sphere f is dependent upon the effective refractive index $N = n/n_o$ (n and n_o are refractive indices of the sphere and the surrounding medium respectively), and it decreases with increasing n.

For this reason, highly refractive glasses with n up to 2.3 are used in retroreflective foils. Such glasses are made from either

Optical and ophthalmic glass 167

lead-alkali-silicates or more chemically stable and lead-free barium-titanium-zirconium-borosilicate compounds.

Manufacturing process

The industrial manufacturing methods for glass spheres depend on the glass composition and the desired sphere size. For microspheres (diameter ≤ 0.1 mm) made of low melting calcium-alkali-borosilicate or lead-alkali-silicate glasses (refractive index 1.5 to 1.65), the following manufacturing methods are used:

- The gravity shaft process is the simplest for easy to melt glass types. Presifted glass particles crushed to approximately the desired sphere size trickle through a heated shaft, where the particles melt and assume a spherical shape. The spheres are collected in a container below the shaft.
- In another process, presifted glass crushed to the desired particle size is fed into a furnace which has an upward directed flame. Under the influence of the temperature and the surface tension, the particles assume spherical shape in the flame. The hot air stream carries them upwards into a cooler zone, where the spheres solidify quickly and are collected in a suitable container.
- The high melting point barium-titanium glasses and the high lead content glasses with high refractive indices crystallize easily. Therefore, cooling time between formation and solidification of the spheres must be very short (quenching at about 1000°C/s). This can be done, for instance, by spraying a thin stream of molten glass under high pressure into a collecting chamber.

In the plastic industry, injection molded and extruded parts (e.g., acetal polymers) are reinforced with up to 30% microspheres. This produces warp-resistant parts with very high rigidity, hardness, form stability, better surface hardness and thermal stability as well as excellent sliding behavior.

Another application for glass microspheres is the micro-blast technique, i.e. similar to sandblasting, microspheres are projected under 4–6 bars pressure onto parts to be fine-lapped or precision-deburred. The microspheres act like micro dropballs, producing a polishing effect and creating a peculiarly glossy surface.

6.7 GLASS FIBER

Glass fiber is a collective term for glass processed into fibers with diameters of between one tenth and a few thousandths of a millimeter. The development of techniques for drawing fibers from a molten glass mass has opened up a variety of new applications for glass.

For some time, two main groups have been separately pursued – insulating fiberglass and textile fiberglass. Both have different applications and characteristics and are made of different types of glass.

A more recent development is the lightguide fiber or optical fiber. These have an increasing variety of potential applications in science and technology.

6.7.1 Insulating glass fibers

These are normally made of soda-lime glass in a centrifugal process using rotating disks. Glass droplets are drawn into fibers of a certain length by centrifugal force (rotation). The fibers cool in air and solidify. The same effect can be achieved in an airblast process, in which air jets cause the formation of fibers from molten glass. The individual fibers immediately mat together to what is sometimes called glass wadding which can either be used as it is or further processed.

The weight of loosely piled, insulating glass fibers is between 30 and 200 kilograms per cubic meter. This makes them particularly suitable for building insulation, since they add no significant load to the structure. Fiberglass insulation can easily be combined with other construction materials, such as mortar and plaster. The durability of the glass ensures a long life for fiberglass insulating products.

Products from insulating glass fibers

Glass wool, another name for insulating fiberglass, is used in the form of loose glass wadding, or as strips, mats, felts, panels, dishes and pressed parts. The fiberglass is either impregnated with a resin or sandwiched between sheets of paper. Dishes and otherwise-shaped parts are formed from resin-impregnated glass wool. Soft matting or stiff panels are pressed.

Uses of fiberglass insulation

In construction, especially residential construction, fiberglass is used mainly for thermal and acoustic insulation. The mats or panels are applied in exterior walls while fiberglass wool is used in partition walls or beneath floors leading to considerable savings in heating costs. In the same way, hot water pipes are insulated against the environment. The soundproofing ability of fiberglass products is due to the many small hollow spaces which impede the propagation of sound waves. Fiberglass insulation is an indispensable product for modern energy savings and environmental protection.

6.7.2 Fiberglass textiles

These are also made of fine fibers produced from molten glass. The fibers are uniform, usually have a circular cross section, and can be further processed into threads. The diameter of the fibers is less than 18 μm. They are subdivided according to length and manufacturing method into glass filaments and glass staple fibers.

Glass silk

This consists of individual glass fibers (glass filaments) with diameters between 5 and 18 μm and any given length that are spun into glass threads for final use.

To manufacture glass fiber-filament yarn, a specific raw material batch is melted in a tank. The molten glass flows in conduits, from where it is downward drawn at high speed into very fine threads (filaments) through an array of platinum nozzles and then rolled onto a drum.

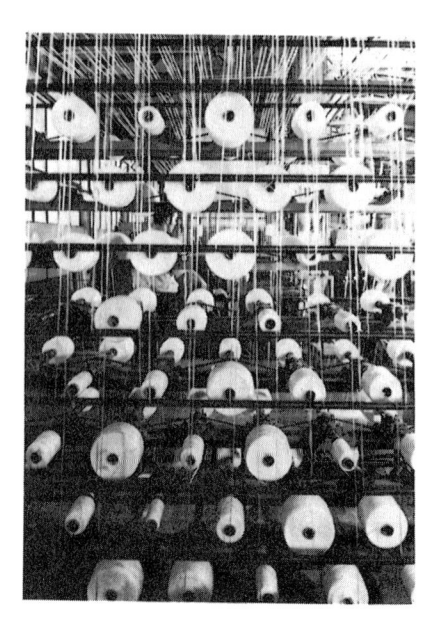

Fig. 6.30 Spool rack for glass staple fiber yarn.

Glass threads

Before the glass fibers can be reeled up on a drum into a 'cake', the filaments are arranged in up to 50 filaments or multiples thereof in order to form a thread. Next, the threads are spooled off the cake and formed into a so-called spun roving as a trade product (Fig. 6.30). The tensile strength of freshly drawn glass fibers is as much as 50% of the theoretical strength of defect-free glass (General characteristics, Section 2.2). Several drawn threads are twined together into glass silk yarn. Glass silk twine is made by winding several glass silk yarns together.

Glass staple fibers

These are glass textile fibers of a certain length and diameter. In an airblast process the molten glass flows through platinum nozzles in the bottom of a platinum melting tank. Compressed air directed over the nozzle opening blows the glass into filaments between 3 and 60 mm long. They are sucked through a perforated drum from which they are drawn into glass staple fibers.

Owing to the manufacturing process and the partially crossed fibers, these fibers are only half as strong as drawn glass fibers. However, because their manufacture is less costly, they are preferred to glass silk yarn for certain applications.

Glass types used in making fiberglass textiles

The most widely used type is the so-called E-glass. It is essentially alkali-free (sodium and potassium content less than 0.8%) and is characterized by its water resistance and high softening point. E-glass is an alumina-borosilicate glass by composition. If acid resistance is required, S-glass is used. This is an alkali-lime glass with a high content of boron and is primarily utilized in woven glass fiber cloth for corrosion protection, gaskets, and surface treatment. D-glass is used where dielectric qualities are important; R-glass, for more stringent mechanical requirements. M-glass has a high elasticity modulus.

Processing additives

Untreated glass threads and glass staple fibers cannot easily be processed because of their surface sensitivity. Hence, certain treatments are applied to their surfaces before further processing. Dressings and lubricants are organic substances with which the glass fibers are coated immediately after being drawn. This ensures that the scouring effect of glass-to-glass contact is minimal and the danger of mechanical damage which could occur during spinning is limited. In drawn glass fibers, the additive is called a dressing, in glass staple fibers, a lubricant.

Adhesive agents

These are needed to improve the adhesion between synthetic resins and the textile glass layers in glass fiber reinforced plastics. Adhesive agents are blended with either the resin or the lubricants. They ensure the bonding of both materials, and they reduce shrinkage of the curing resin.

Fig. 6.31 Fiberglass reinforced plastic canoe.

Fiberglass reinforced plastics

High tensile strength of up to $10^3 \, N/mm^2$ and low expansion are the properties which render fiberglass suitable for improving the mechanical qualities of plastics. Moreover, the strength of fiberglass reinforced plastics increases as the glass content increases. The fibers are layered into the plastic in the form of threads, bands, strings or cloth. Embedded in the plastic, glass fibers assume the mechanical load placed on the material before it is transferred to the plastic itself. Impact resistance and high vibration absorption make fiberglass reinforced plastics the ideal material for protective helmets, chassis parts, telephone booths, loudspeaker housings, etc. (Fig. 6.31). Colors are virtually unlimited and because of the excellent weathering characteristics surfaces do not need to be lacquered. By controlling the difference between the refractive indices of the fibers and synthetic resin, transparent or translucent skylights, lighting fixtures or storage tanks with visible level indication can be made. The material is also resistant to many chemicals, so that it can be used to make acid containers, plating troughs, piping, fodder silos, and even beverage and food containers.

Textile glass fiber

When impregnated by polytetrafluoroethylene (PTFE), this has non-stick properties, is resistant to chemicals and heat, is nearly inflammable and has a high mechanical resistance and electrical insulating properties.

Glass fiber reinforced concrete

Adding up to 5% glass fibers to the concrete mix results in an increased flexural and tensile strength, better impact resistance and breaking stress behavior as well as a greatly reduced tendency to cracking as compared with ordinary concrete. Owing to an almost unlimited resistance to weathering and other environmental influences, glass fiber reinforced concrete panels are extremely well suited for facades. Glass fibers with a high alkali resistance are required for glass fiber-reinforced cement concrete due to their low weight. High performance laminated rods of glass fibers and polyester resins with a high elastic extension and a tensile strength comparable to prestressed steel have proven extremely useful in the construction of prestressed-concrete bridges.

Glass fiber-reinforced glass

The Research Department of Schott is in the process of developing a breakage resistant glass by bonding glass fibers, 0.01 mm thick, with glass powder under pressure and a temperature of approximately 1350°C. The composite material shows a higher breakage resistance than steel at a much lower weight. Its stability under thermal stress makes it particularly interesting for the manufacture of engine and aircraft components.

6.7.3 Glass fiber optics

This term covers a variety of high-quality products which are made with optical fibers.

Fiber light guides

Fiber light guides are optical systems capable of propagating light in any chosen direction – around bends, curves and loops (Fig. 6.32). A fiber light guide is a thin, flexible filament with a diameter of a few hundredths of a millimeter. The interior or core consists of high refractive index optical glass surrounded by a sheath of glass of a lower refractive index. If a light beam enters at one end of a fiber with an incident angle that is

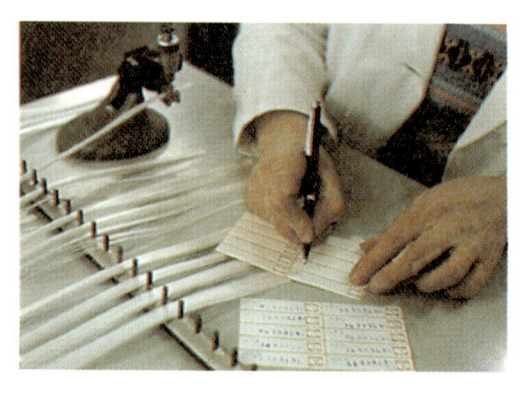

Fig. 6.32 Fiber optic light guides.

defined by the difference between the refractive indices of the core and sheath glass, it travels by total reflection at the core/sheath interface to the opposite end of the fiber. In a fiber bundle, the sheath optically insulates the cores from one another, so that light from one fiber cannot be transmitted to an adjacent fiber. Optical fibers which operate according to this principle are also called step index fibers.

The manufacture of light guide fibers

In the rod-tube process, a rod of high refractive index glass is placed inside a tube of low refractive index glass. They are then heated in a furnace where both glasses soften simultaneously and are drawn to thin fibers that are wound on drums or spools. In the two-crucible process, both glasses are melted separately before being drawn through a concentric dual orifice into a fiber and then wound on reels (Fig. 6.33). It is also possible to produce glass fibers with the sol-gel method (Section 3.3.8).

Light guides

These are bundles of light guide fibers which are either completely fused to each other over their total length (rigid light guides) or only fused at their ends (flexible light guides) (Fig. 6.34). The ends are ground and polished perpendicularly

Glass fiber 175

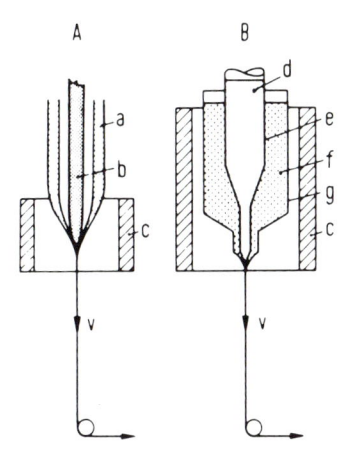

Fig. 6.33 Drawing of light guide filters from core glass and sheathing glass: (A) with the rod/tube process; (B) with the two crucible pots process.
(a) Glass tube, (b) glass rod, (c) ring heater, (d) core material melt, (e) inner crucible, (f) sheath material melt, (g) outer crucible.

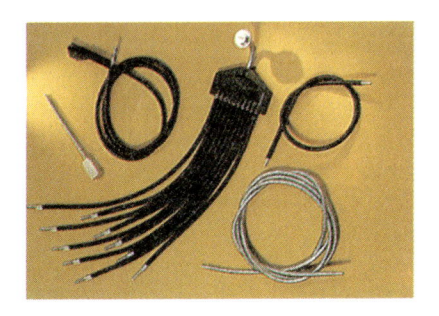

Fig. 6.34 Light guide cables with glass filters.

to the fiber axis. In order to protect them from environmental influences, the fiber bundles are enclosed in flexible casing or rigid housings.

Image guide

In image guides the optical fibers maintain the same position on both ends of the bundle, so that images are picked up as a

Fig. 6.35 Rigid image guide made of glass fiber light guides.

multitude of individual points at one end and transmitted to the same spot at the other end. Each fiber transmits only one point of the image. The smaller the diameter of the fibers, the higher the resolution, i.e. the separation of neighboring image points (Fig. 6.35).

Light guide cables

By jacketing fiber bundles with plastics, continuous light guide cables can be made and reeled on drums. They are basically assembled like electrical cables. Shorter cables are sometimes sheathed in flexible metallic casing instead of plastic. For cables with a desired length, the ends are packed as densely as possible, then embedded into a metallic ferrule.

Multiple branch light guide cables can be used to either distribute light from a single source to many points, or concentrate light from many sources at one point. The two ends of the fiber bundles can also have different cross sections which then results in a change of the luminous flux distribution.

The uses of glass fiber optics

Light guides are used where normal heat generating light sources cannot be directly used, for instance in the medical field of endoscopy, where only cold light can be used to illuminate internal human organs. Cold light is obtained by filtering out the heat radiating components of light before they enter the light guide. Flexible fiber bundles are also used in

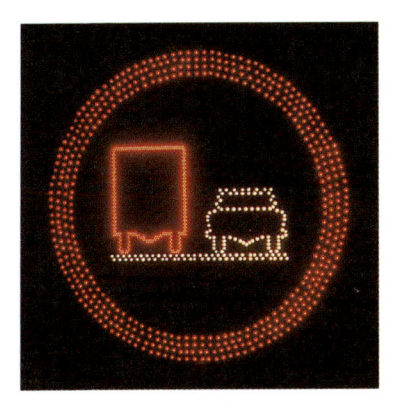

Fig. 6.36 Changeable traffic message sign using light guide fibers.

microscopy and for close precision work illumination, e.g., in the restoration of ancient art works. The reading heads in computers are built with fiber optic light guides, and so are control panels, changeable message traffic signs and oil furnaces (the light of the flame is transmitted to a shielded photocell). In changeable message traffic signs, a large number of single fiber cables is assembled to a multi-branch harness whose common end is illuminated by one central light source. The various branches are then arranged to form letters or numbers on a display panel (Fig. 6.36).

Ultraviolet transmitting light guides, with quartz glass cores and low refraction plastic sheaths, are used in dentistry for caries treatment.

Fused fiber optic plates made of fibers of approximately 5–10 μm diameter and just a few millimeters in length can be used as image field flatteners (i.e. to correct the deviations of the image from the focal plane in order to obtain a clean, sharp two-dimensional image in the focal plane). Image converters and low-light-level amplifier tubes using fiber optic microchannel plates have greatly improved night-vision technology.

Fiber light guides used in communications technology

The most important use of optical fibers in the future will undoubtedly be in communications technology. They will respond to the rapidly increasing demand for more capacity for telephone and television transmissions while at the same time replacing copper that is falling into short supply worldwide. The electrical impulses of the transmitting station are converted by semiconductor devices into light (or infrared) signals and transmitted over great distances without intermediate amplification. At the receiving end, photodiodes and amplifiers again transform the light signals into electrical signals, which then operate the telephone or television set. A prerequisite for this application of fibers was a drastic reduction in optical losses in the glass. While the light intensity falls to 50% after travelling over only one meter in standard quality optical glass, special materials and processes have reduced the attenuation in optical fibers (for wavelengths between 0.8 and 1.6 μm) to a value that requires intermediate amplifiers only every 100 km. This is equivalent to a loss of about 0.3 dB/km. In order to obtain a transmission capacity of more than 1 gigabit × km/s (1 bit is one element of information (0 or 1); 1 gigabit = 10^9 bits) – which is needed to simultaneously transmit 20 000 telephone calls or 20 TV channels – graded index fibers are necessary (Properties of optical glasses, Section 6.6.1). Graded index fibers have a refractive index falling from the core center to the edges of the fiber following a predetermined mathematical function, thereby offering an almost uniform optical path length for all modes, i.e. for all light rays irrespective of their angle to the fiber axis. Such systems have already been successfully installed (Fig. 6.37).

In a step index fiber, light incident at oblique axis is tracing a zig-zag path through the fibers travelling a longer distance than light going straight down the fiber center. That is why not all light arrives at the same time at the far end and the fiber is said to have a high signal dispersion. In a graded index fiber, light rays travelling away from the central axis pass through zones with progressively lower refractive indices. Since the velocity of light in a medium is inversely proportional to the refractive index, rays travelling in low gradient zones are faster than those travelling down the center of the fiber. When the right refractive index gradient is chosen, the net result is that virtually all light arrives at almost the same time at the far end, and signal dispersion is greatly reduced. Graded index fibers are presently the material of choice for telecommunication fibers.

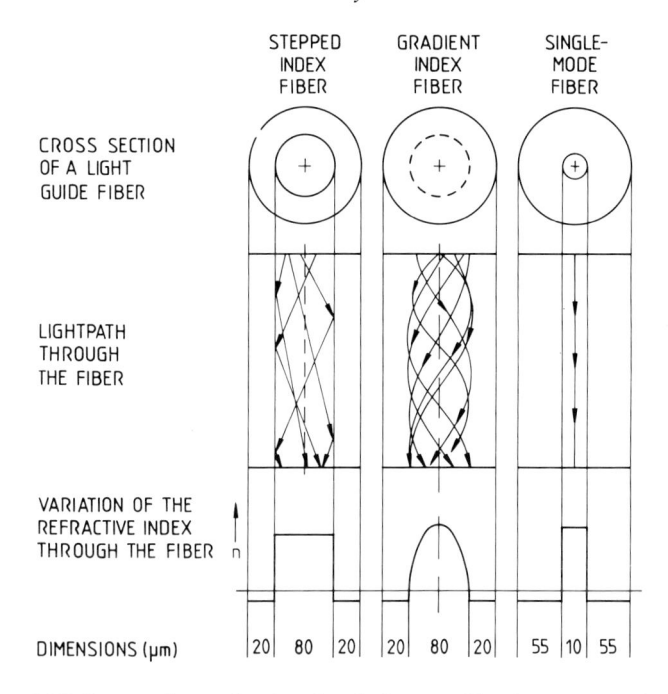

Fig. 6.37 Comparison of refractive index profiles in a stepped index fiber, gradient index fiber and a single mode fiber.

This will probably change in the future, however. Owing to the improvement of fiber installation and alignment techniques in recent years, single mode fibers have become more easily adaptable for applications and will soon replace graded index (multi-mode) fibers.

A prototype single mode fiber consists of a cladded homogeneous fiber core of only a few micrometers diameter. This small size is responsible both for the difficult handling of single-mode fibers and for their superior transmission capacity. Because of this geometry, only rays travelling in the direction of the fiber axis can propagate. All of the other rays lose their whole power by lateral radiation after a short distance, and thus are effectively cut off. This effect is due to the wave nature of light, which must always be taken into account if the dimension involved, in this case the core diameter, has the same order of magnitude as the wavelength.

As a result, there is only one way or 'mode' for the light to

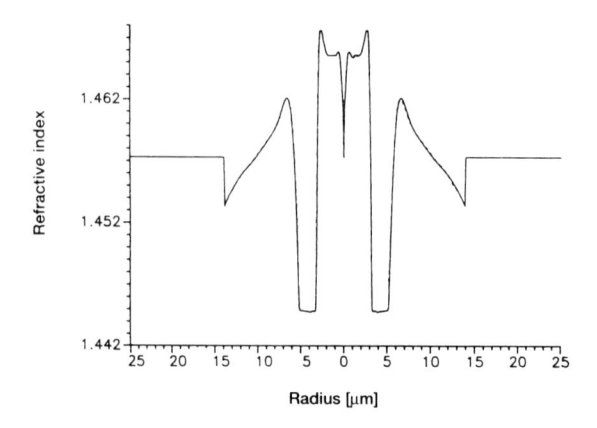

Fig. 6.38 Refractive index profile of a CP^2 fiber.

propagate in 'single mode fibers'. They can show no signal dispersion because of different ray paths, in contrast to the graded index (multi-mode) fibers, where this dispersion effect is greatly reduced but not completely eliminated.

The system performance, however, is also limited with single mode fibers because the velocity of light depends upon the wavelength in refractive materials. If a light source with a non-vanishing spectral width is used, this so-called chromatic dispersion gives rise to a broadening of the light pulse and, as a consequence, limits the maximum pulse sequence.

Nevertheless, the wavelength dependence of the velocity of light can be influenced by the waveguide structure (by the refractive index profile). Schott's major contribution to such low chromatic dispersion single mode fibers has been the CP^2 fiber, where an optimum performance has been achieved with continuous refractive index profiling (Fig. 6.38).

Using a CP^2 fiber and a conventional light-emitting diode (LED) as a light source, a transmission capacity of approximately 1 terabit × km/s (1 terabit = 10^{12} bits) has been reached at both 1.3 μm and at 1.55 μm. At these two wavelengths, quartz fibers reach their two lowest absorption minima. Standard single mode fibers have low dispersion values only at either one of these wavelengths, not both. Continuous profiling, therefore, has resulted in a fiber that is compatible with both standard types and thus has been named CP^2.

The desired refractive index profile in a graded index

Glass fiber

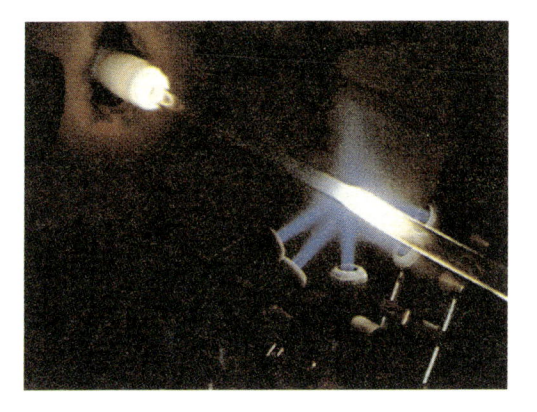

Fig. 6.39 Manufacture of the preform for graded fibers.

multi-mode fiber, as well as in a continuous profile single mode fiber, is produced by heating a quartz glass tube to a very high temperature and passing doses of vaporized compounds of glass-forming elements (Si, F, B, Ge, P) together with oxygen through the fiber. The resulting glassy oxides precipitate on the hot inner wall of the tube in a preprogrammed mixture ratio. Then the tube is heated until it collapses. The fiber is drawn from the resulting 'preform', maintaining the same refractive index profile inside the solid fiber, only on a smaller scale (Fig. 6.39).

In a recently developed process, glass fibers can now be assembled and glued together to form a ribbon cable which is then placed in a plastic tube. A ribbon cable is about half as thick as a stranded cable.

High transmission capacity, small space requirements, and the total insensitivity to external electromagnetic interference make optical communication fibers an ideal transmission medium for modern telecommunication. In the summer of 1992 a transatlantic glass fiber cable with a capacity of 60 000 telephone channels between the USA and Germany was put in operation. Over the total length of 7200 km (4500 miles) between Green Hill/Rhode Island and Norden, Germany, 56 repeaters had to be installed for intermediate amplification of the signals.

In optical fiber networks for TV transmissions, such electronic repeaters are necessary every 130 km (80 miles) in order to compensate for the cable losses. In the future, these electronic repeaters will be replaced by optical travelling wave amplifiers. That would increase the distance between two repeaters to approximately 1000 km (625 miles).

In recently conducted promising tests in the Bell Laboratories (AT & T) a pumping light beam supplying the energy for these amplifiers was fed into the fiber cable together with the information-carrying pulses. The goal is to prevent the distortion of the information pulses, caused by the dispersion in the fiber over the long distances, using stable pulses (solitons) that do not undergo any distortion.

Optical fibers for measuring and control applications

In addition to their application in telecommunications, low loss optical fibers have led to a variety of innovations in such areas as operational and functional control, remote control and remote detection and sensing. One example is the optical triggering of thyristors in equipment for high voltage-DC transmission from remote hydro-power plants to industrial zones. The electrical control signal is converted via semiconductor diodes into a 0.94 μm infrared signal which is transmitted to the thyristor, connected to high potential, via an infrared-light conductor. Reversed into an electrical signal, it fires the thyristor. An analog revertive data link to the control unit monitors the function of the system. This installation is guaranteed to operate without any interference even at nominal voltages of 600 kV and power transmission of 3000 MW.

Optical fiber cables are also used as sensors in the technical diagnosis of engines and heavy machinery where the measuring or observation area is too far away or difficult to be reached directly. For instance, this allows a continuous surveillance of the inside of turbines and compressors as well as the real time measurement of vibrations, temperatures and pressure values. Movements of the sensor fiber of less than $\lambda/2$ (λ is the optical wavelength in use) are still detectable because of the change of the optical path length in interferometer–fiber circuits.

Fissures in structural components, such as aircraft wings, may be detected using optical glass fibers embedded into the surface finishing of the component. If the glass fiber is damaged by a crack in the component, a signal is automatically transmitted to the control unit. The same effect is exploited for the safety of vault rooms or in the open air where light guides are used in the form of mats or wired together to detect surface damage or warping. Any damage to the light guides automatically activates an alarm system.

Integrated optics

Light guides can be manufactured not only in the form of thin glass fibers, but also in flat glass substrates as extremely fine structures. As the resulting optical components are monolithic light guide structures with often complex optical functions, they are also called integrated optical circuits (in analogy to the electronic integrated circuits). The basic elements are integrated optical chips in glass in which the light guide function is generated by ion diffusion using a mask. The mask determines the light guide structure in the substrate glass. It is created as a metallic vapor deposit on the substrate surface into which the mask pattern is etched using a photolithographic process, known from the semiconductor industry. In the following ion exchange process, the sodium ions of the applied substrate glass are replaced by silver ions thus creating areas of a higher refractive index wherever glass is exposed to that process. These areas then function as light guides.

Examples of such integrated optics components are the splitters (1:4, 1:8, 1:16) that are required in fiber optic transmission networks, or the 3×3 coupler as used in a miniaturized Michelson interferometer with integrated optical signal-preprocessor.

6.8 GLASS-CERAMICS

As already mentioned in Chapter 2, all glasses are in the state of a super-cooled liquid when the melt is cooled below the melting point of crystals of the same chemical composition. Crystallization (devitrification) does not occur because crystal growth, which is controlled by the diffusion of the components, is either much too slow, as a result of the rapidly increasing viscosity of the molten glass with the falling temperature, or because the number of nuclei, from which crystallites (smallest particles identifiable as crystals) can be formed, is too

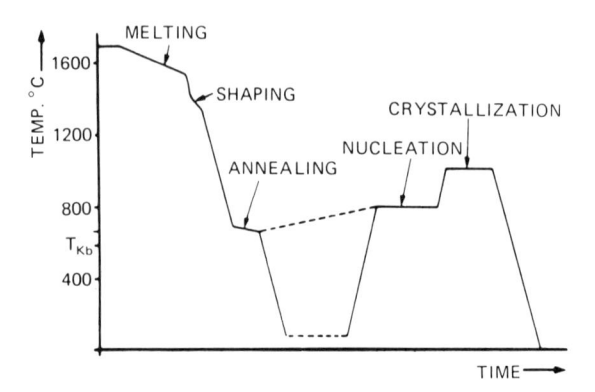

Fig. 6.40 Basic temperature vs. time profile of a glass-ceramic ceramizing process.

low. In glass-ceramics, on the other hand, crystalline growth is deliberately stimulated in appropriate glass systems in order to obtain materials with special properties.

The starting point is a melt of appropriate glass components from which the desired items are formed by pressing, blowing, rolling or casting. One special technique worth mentioning is the centrifugal cast process developed by Schott for the manufacture of telescope mirror blanks of 8.2 m (27 ft) diameter and only 30 cm (12 in) thick, with a total weight of 40 metric tons (88 000 lb).

During a subsequent heat treatment following a defined temperature–time profile (Fig. 6.40), submicroscopic crystallites begin to form provided that high melting point materials (usually TiO_2 and ZrO_2) have been added to the melt where they precipitate and act as nuclei to set the crystallization in motion. It is important that the temperature zone of the maximum nucleation rate (T_{Kb}) lies far enough from the temperature zone of maximum crystal growth rate (T_{Kr}) (Fig. 6.41). The molten glass simply cannot crystallize during the cool-down period as long as there are no nuclei present in the melt. Only when a sufficient number of nuclei have formed at T_{Kb} can the minute crystallites be formed into large quantities (up to $10^{17}/cm^3$) when the glass is reheated to T_{Kr}. The crystallized portion in the glass-ceramic by volume can be between 50% and 90%, depending on the desired properties.

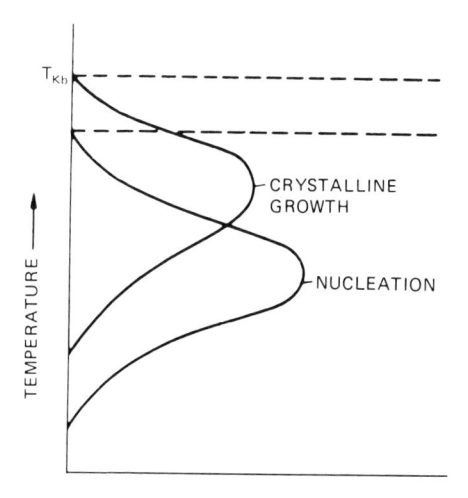

Fig. 6.41 Nucleation and crystalline growth rates vs. temperature.

The technical importance of glass-ceramics is characterized by their properties. These are not only determined by their glass content, but also by the types of crystals they contain. Some systems, which have crystalline phases with very low or even negative thermal expansion (such as lithium-alumina-silicates), have become very important. They form materials with near-zero expansion over a wide temperature range, maintaining their shape up to 800°C and being totally insensitive to thermal shock. They are used for cook-top panels, cookware, mirror substrates for telescopes, standards gauges, etc. (Fig. 6.42).

In another group of glass-ceramics built with the components Si, Al, Mg, K, F and O, mica-like crystals form during the 'ceramizing' process that can easily be split and allow the glass-ceramic to be machined (e.g., on lathes) (*Macor*, Corning). Glass-ceramic of that kind is also used as a highly resistant material for dental restoration and for tooth crowns in true to nature shape and color.

Ranging somewhere between glasses and ceramics, there are also various photosensitive glass types where the crystallization can be initiated by ultraviolet radiation and then be finalized by a heating process. For this to occur, certain amounts of alkali fluoride, zinc and aluminum oxides, as well

Fig. 6.42 Telescope mirror substrate blank made of *Zerodur* glass-ceramic, 8.2 m ϕ.

as small amounts of silver compounds ($\leq 0.05\%$) in the silicate glass matrix must be present. Ultraviolet radiation causes electrons to separate from the cerium ions which are then trapped by the silver ions when heated, thus forming metal atoms which immediately transform into metal colloid particles. They act as nuclei for the devitrification process. When further heated, this structure transforms into a yellow or brown glass-ceramic. In this way, ultraviolet photographs can be stored and fixed in the glass. It is also noteworthy that the crystallites occurring in lithium silicate glasses can be etched out with hydrofluoric acid at least ten times faster than the surrounding glass matrix. This property facilitates the production of high-precision etched parts such as perforated plates for displays, printing dies and the like (Fig. 6.43). The presence of other halides (bromine and chlorine) produces various other colors in the glass. These new polychromatic glasses (Corning) will find technical as well as decorative use in the future.

6.9 POROUS GLASS AND FOAM GLASS

In the description of the batch melting process (Melting process, Section 3.3), it was mentioned that the primary melt

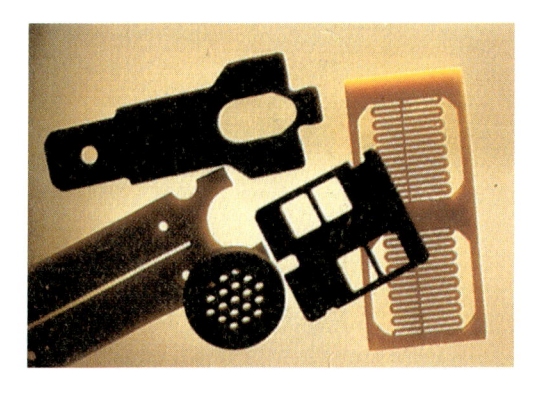

Fig. 6.43 Chemically etched components made of photosensitive glass-ceramic.

has a high gas content which would inevitably result in a glass with a high bubble content if the melt were not adequately refined. If the opposite approach is used, and additional gases or gas forming components are deliberately added to the melt, the quantity of bubbles can be increased to produce foam glass.

In another process a mixture of powdered soda-lime glass, pulverized sulfates and charcoal is filled into molds and sintered. When this mixture melts it forms SO_2 and CO gas. Under the influence of increasing gas pressure the initially spherical gas bubbles assume a honeycomb structure containing polyhedral cells. The result is a foam glass with high compressive strength and dimensional stability with a density of only 0.13 to 0.3 g/cm^3. The cellular structure reduces the thermal conductivity of the product to about one tenth of that of a compact glass. These properties make foam glass particularly suitable for thermally and acoustically insulating construction materials. Foam glass can also be coated with concrete to be used as flooring material. Owing to its low density and its water impermeability, it is extremely well-suited for flotation devices.

Open-pored glasses and glass capillary membranes

Foam glass has closed pores. Recently developed porous glasses and glass-ceramics on the other hand have open pores which to some degree extend right through the material. These are produced in defined sizes by the conventional sinter process used for ceramics. This involves the formation of openings by blending in organic fillers which are subsequently burned out. The resulting material has a minimum bulk density of 0.3 g/cm^3 and a void volume in the region of 90%, which gives it a high absorption and storage capacity (like a sort of glass sponge) and in view of its great chemical and mechanical durability, it provides the possibility of storing and transporting hazardous fluids safely.

Sodium borosilicate glass with a 20–70% SiO_2 content is particularly suitable for the production of microporous glasses (capillary membranes) with pore diameters of 2–300 nm. In these glasses a phase separation occurs between 500 and 750°C forming an alkali-borate rich phase that can easily be leached out (Fused silica, Section 6.1). The pore structure produced in this way creates a specific surface area of up to 300 m^2/g on which silanol (SiOH) groups from the glass itself are immediately formed; functional groups of various types can be chemically bonded to these by reaction. It is expected that applications for glass membranes will be found in sea water desalination plants and in the health-care sector in areas such as blood treatment or artificial kidneys.

6.10 A GLANCE INTO THE FUTURE

The increasingly better knowledge of and the successful developments in glass technology over the past few years suggest that the possibilities of finding glasses with special properties for new applications or finding new technologies to make such glasses are far from exhausted. Indications of this were described in 'Electron conductive glasses' in Section 6.4.8. Several years ago, it was discovered that certain oxide-free metallic compounds can be converted into a glassy state if quenched extremely rapidly (10^5 K/s). Such cooling rates can only be achieved with thin ribbons or wires. In the USA, several alloys are already being processed into narrow glassy ribbons in a continuous casting process. Precious metals and transitional metals, such as Pd, Fe, Ni and Cr combined with one or more metalloids (Si, P, B, C) are especially suitable for this. Properties that are typically metallic such as flexibility, high strength, electrical conductivity, and even ferromagnetism are maintained despite the amorphous (noncrystalline) structure of the products. In another approach it was discovered that an originally opaque sandwich of two transparent flat glass panels with a liquid crystal foil in between becomes transparent when connected to an electrical source. The electrical field between the indium oxide coating of the two glass panels orients the randomly oriented crystals in the foil, making the opaque window transparent. When the voltage is cut off it goes back to its original opaque state. Another goal of intensive research is the reduction of the brittleness of glass, which would open up a wide range of new applications for glass.

Promising approaches to new glass manufacturing techniques are also being conducted with the objective of producing glass without a high energy-consuming melting process. One promising method uses silica gel solutions with additions of soluble compounds of metals or hydrolyzable alkoxides of several glass forming elements. These are slowly heated and brought to reaction and polycondensation by splitting off water. This principle has long been in use to produce special thin film dip coatings on glass, but no viable process yet exists for making larger glass components. Success in this direction could initiate a great step forward in modern glass technology.

7
Environmental protection in the glass melting process

Melting and processing glass is to a great extent done in processes generating by-products that are released into the air or into water or become scrap or waste material polluting the environment. Very often they are also hazardous; however, there are specifically developed treatments and techniques which substantially reduce or eliminate their environmental impact. Legal regulations allow the government to influence and control the necessary measures and ways to assure their application.

7.1 GLASS MELTING

Glass melting is basically a high-temperature process creating a multitude of emissions of various importance, described below.

7.1.1 Solid particle emissions

Replenish the raw material silos and preparing the glass batch are dust creating processes, as is the batch loading of the tanks. The most important source of particles, however, is the glass melt itself where volatile metal compounds condense in cooler spots and areas of the equipment. They react partially with compounds from flue gases forming new solid compounds. In this way, oxides, i.e. sulfur oxides and various halogenides of alkalis, are generated. In special glass melts additional compounds are formed with lead, boron, arsenic, antimony, cadmium, zinc, etc.

7.1.2 Gaseous emissions

In addition to the moisture that is released from the raw material batch during the melting process, other gaseous compounds result from hydrates, carbonates, nitrates, sulfates, fluorides and chlorides of the various batch components. They are released in the form of water vapor H_2O, carbon dioxide CO_2, nitrogen oxides NO_x, sulfur oxides SO_x, hydrofluoric acid HF or silicontetrafluoride SiF_4 and hydrochloric acid HCl. Another important contribution to the gaseous emissions in the process comes from the combustion of fossil energy, such as natural gas and fuel oil, i.e. water vapor, carbon dioxide and sulfur oxides as well as nitrogen oxides formed by oxidation of atmospheric nitrogen.

7.1.3 Flue gas dust collection

In areas where dust develops at low temperatures, i.e. raw material storage and batch preparation, the cleaning is done by separating filter systems. By contrast, the mixture of dust and gases from the melting tank is at a substantially higher temperature level, even after the heat exchanger. The dust is collected either by filtering systems or electrostatically. In some cases wet cleaning is necessary.

Special attention is given to the pH-value (acidity) of the gas–dust mixture; if it is low due to a greater portion of acids and acid anhydrides (SO_2, SO_3, HCl, HF, etc.) it becomes chemically aggressive and harmful to the equipment. In order to prevent such damage, alkaline compounds such as calcium hydrate or sodium carbonate have to be mixed into the raw gas flow to bind with and neutralize the acidic components. The resulting compounds are basically calcium sulfite/sulfate $CaSO_3/CaSO_4$, calcium fluoride CaF_2, sodium chloride NaCl and other compounds that can be separated and filtered out as small particles. A dust–gas mixture can be cleaned to a great extent using this well-proven technique.

Several chemical reduction techniques have been tested to eliminate the nitrogen oxides. According to the reaction process:

$$4\,NO + 4\,NH_3 + O_2 \rightarrow 4\,N_2 + 6\,H_2O$$

the addition of ammonia or other amine compounds generates nitrogen and water vapor. The reaction process can take place either directly at temperature levels of 850 to 1100°C or via catalysts at temperatures between 250 and 450°C (e.g., using ceramic honeycomb modules in titanium oxide, TiO_2). This technique has been tested in its principal application, and its introduction on an industrial level has already begun.

The use of fully electrically heated glass melting tanks combined with the cold-top-technology is a promising way to reduce substantially the volume of local emissions. This could become the process of the future for certain applications.

The highly desired reduction of carbon dioxide emissions – a major factor influencing our climate – can only be achieved by increased energy saving efforts.

7.2 WASTE DISPOSAL

The various steps in the melting and processing of glass create by-products that were traditionally disposed of as waste; these include batch residues, glass cullets and broken glass of a given composition, filter dust, chamber condensation products, tank liner fragments, grinding and polishing sludge, neutralization sludge of various processes, as well as residues from refinement and finishing processes and many more.

Today's approach is to put first priority on efforts to avoid the formation of these by-products. Where this is not or not yet feasible, recycling and reuse of the by-products has absolute priority over the final disposal in dumping grounds. This strategy is also fuelled by the tremendous cost increase in all disposal options.

Avoiding the formation of by-products requires, in general, major changes in the manufacturing process and can only be successful in the long term. In contrast, recycling and reuse of by-products has accelerated. Filter dust, glass cullets (used glass recycling), fragments of the tank lining and polishing sludge can already be recovered and recycled to a great extent, either in process or externally off site. Other products have to be processed at high cost and thus, still present an economic hurdle at this time. Finally, some by-products still have to be disposed of in designated dumping grounds.

Other industrial waste, such as packaging material of all

kinds, can be sorted and almost entirely recycled. Other materials, such as TV tubes or *Ceran* cook-top panels, are still the subject of an intensive search for appropriate recycling methods. Behind all these efforts lies the vision of an economy that works in cycles with a minimum drainage on the world's decreasing and irreplaceable resources.

8
Glass as an economic factor

In all industrialized countries, the glass industry is one of the smaller players in the entire production industry. At most, its value is one percent of total production.

In 1992, world glass production was estimated at 90 billion DEM with an annual output of about 85 million tons. The single largest glass-producing nation is the USA (over 25%). The Western European countries produce 32%, Asian countries about 20%. Of the total value of world glass production, flat glass accounts for about 35%, hollowware 45%, special glass 10% and fiberglass 10%.

The 290 or so glass-producing, finishing, and processing firms in Germany employed approximately 70 000 people in 1982. Germany exports an average of 30% of its total glass production. Some export-intensive glass branches export a greater proportion of their output; such as tableware and special glass, with more than 50% of their total output exported to over 100 countries. The produce range of the German glass industry is matched in its comprehensiveness only by those of the USA and Japan. It has a key position in the German economy. The glass industry has very close supplier relationships to many large industries. Glass has become an important component of many of their own products. At the top of the list of glass consumers is the construction industry. It is followed closely by the food and beverage industry. Other consumer groups include households, catering, the automobile industry, the electrical industry, plastics and textiles, chemical and pharmaceutical industry, medicine and research, furniture industry and the optical industry.

Historically, the European countries are the classical glass producers. When the Industrial Revolution began in the second half of the eighteenth[1] century, it also embraced glass production which had hitherto been entirely made by craftsmen's experience. The combined glass industries of the European Community now come close to the production potential of that in the USA. There is also extensive glass production in the East European countries.

Developing nations are also increasing their glass production capacities, usually with the financial and technical support from industrialized nations. In so doing, they create jobs and supply domestic needs for such glass products that are relatively easy to produce. The availability of raw materials for glass almost all over the world encourages this trend.

Appendix

GLASS MUSEUMS

Most art history museums show more or less important glass collections. In addition the following museums give an overview of glass history, of the evolution of manufacturing processes from the past to the present, and of the most important glass products.

Glasmuseum Frauenau	D94258 Frauenau (bei Zwiesel/ Bayer. Wald, Germany)
Glasmuseum Wertheim	D97877 Wertheim/Main, Germany
Kunstsammlungen der Veste Coburg	D96450 Coburg, Germany
Museum of Glass	Corning, New York, 14830, USA
The Toledo Museum of Art	Toledo, Ohio, 43697, USA

EXPLANATION OF PHYSICAL SYMBOLS AND UNITS

Physical and technical data are always expressed as a product of a numerical value and a unit. The fundamental units (meters, seconds, degrees of temperature, etc.) are identified by the symbols m, s, K (kelvin), and so on. The numerical value of data can extend to many powers of ten. Then the units are prefixed by symbols that express factors of several thousands. The most commonly used are:

k (kilo) = 10^3,   M (mega) = 10^6,   G (giga) = 10^9,
m (milli) = 10^{-3}, μ (micro) = 10^{-6},   n (nano) = 10^{-9}.

The often-mentioned unit of length μm therefore means 1/1000 of a millimeter. The most frequently used time units are seconds (s) and hours (h). Frequencies are expressed in Hz (Hertz) = c/s (cycles per second), kHz = 10^3 c/s, MHz = 10^6 c/s, etc.

The absolute temperature is T (in K) = t (in °C) + 273.15. Therefore, temperature differences have the same value in K as in °C. The coefficient of thermal expansion, α, of a solid body is a measure for the relative linear expansion per K (1°C) temperature increase, which in most glasses is in the order of $3-10 \times 10^{-6}$/K.

The technical unit of power is the watt (W), along with kW, MW, etc. Therefore, the unit for energy is the watt second (W s) (also called the joule), and the kilowatt hour (kW h). Because mechanical energy is defined as the product of force and distance, the unit of force N (newton) can also be expressed by W s/m.

A force exerted on a solid body produces, depending on its direction, either tensile or compressive stress, σ, which is measured in N/m^2 = Pa (pascal), and which also serves as a measure of strength determining the maximum allowable load. These units (preferably N/mm^2) are also used to express the modulus of elasticity, E, which is used to determine the relative elastic expansion according to Hooke's law ($\varepsilon = \sigma/E$).

The relationship between the exerted transverse force and the attained speed of a plate moving in a viscous liquid (melt) results in the viscosity coefficient, η, which is expressed in pascal seconds (Pa s); the old unit, poise, is equal to 0.1 Pa s (=1 dPa s).

ATTENUATION OF RADIATION

As radiation (light, X-rays, electrons, etc.) traverses a distance s through matter, it decreases in intensity (J) due to absorption or dispersion, to a value $I_0 e^{-\beta s}$. The material constant β has a dimension of m^{-1}, as βs (= $-\log_e I/I_0$) is a non-dimensional coefficient. It is customary, however, to express such logarithmic ratios in decibel (dB) (= 0.1 bel) for intensity attenuation. Usually, dB-values are in reference to a distance s, whereby for example 1 dB/km corresponds to the value $\beta = 2.303 \times 10^{-1}$ (Fiber light guides, Section 6.7.3). Similarly, dB is used in acoustics to indicate the relative sound intensity (audio volume). Because the human ear perceives sound intensity (P) more or less proportionally to its logarithm, the sound volume level has been defined as $L = 10 \log_{10} P/P_0$, where P is the lowest audible intensity at 1000 Hz ($\approx 10^{-12}$ W/m^2). Thus for example, $P/P_0 = 10^5$ results in a volume of 50 dB. This number corresponds approximately to the phon measurement for the perceived loudness of a noise (Glasses with altered radiation, Section 4.4.1).

TECHNICAL LITERATURE ON GLASS

Jebsen-Marwedel, H. 1976. *Glas in Kultur and Technik.* Selb, Verlag Auman KG.

Klindt, L. B. and Klein, W. 1977. *Glas als Baustoff; Eigenschaften, Anwendungen, Bemessung.* Cologne, Verlagsgesellschaft Rudolf Müller.

Lohmeyer, Sigurd, 1979. *Werkstoff Glas; Sachgerechte Auswahl, optimaler Einsatz, Gestaltung und Pflege.* Grafenau, Expert-Verlag GmbH.

Persson, R. 1969. *Flat Glass Technology.* New York, Plenum.

Pye, L. D. et al. (eds.) 1973. *Introduction to Glass Science.* New York, Plenum.

Scholze, H. 1977. *Glas-Natur; Struktur unde Eigenschaften* (2nd edn). Berlin, Springer Verlag.

Vogel, W. 1979. *Glaschemie* (1st edn). Leipzig, VEB Deutscher Verlag für Grundstoffindustrie.

Index

Page numbers appearing in **bold** refer to figures and page numbers appearing in *italic* refer to tables.

Abbe, Ernst 12, 13
Abbe number 150, 151
Absorption
  of infrared 68, **69**, 135
  of ultraviolet 70, 144, 155, 156
  of X-rays 135
Achromatic lens systems 154
Acid etching 75, 117–18
Acid polishing 116–17
Acoustic insulation 68, 81, 169
Agate glass 114
Air inclusions, *see* Bubbles
Airlines [defects in glass] 49
Alarm glass 79–80
Alkali-borosilicate glasses 131, *132–3*
Alumina [aluminum oxide] 25, 28
Alumina-silicate glass **21**, 131, *132–3*
Annealing points/temperatures *21*, 52, *132*
Antenna glass 81
Antique glass 64–5
Anti-reflection coatings 73
Apochromatic lens systems 154
Apparatus glass 104
Architectural glass products 11, 55, 70–1, 77–8, 108
  see also Window glass
AR-glass 129, *129*
Armor plate glass 79, **80**
Athermalized glasses 153
Attenuation of radiation 68, 197–8

Automotive applications 74, **75**, 77, 79, 146

Barium-borate glasses 148
Barium crown glasses 153
Barium flint glasses 153
Barium oxide 29, 153
Basket bottles 101
Batch 33–4
Batch feeding **40**, 47–8
Bauhaus 10
Beads, glass 10, 98, 99
Beverage bottles 100–1
Bifocal spectacle lenses 157
Birefringence 22
Blisters [in glass] 49
Blow–blow process 89, **91**, 100
Blowing processes
  machine blowing 88–91
  mouth blowing 3, 85–7
Boron compounds 29, 52, 131
Borosilicate glasses 25
  properties **21**, 25, *122*, 123, *132–3*
  uses 51, 63–4, 104–5, **106**, 123–7, 131, 141, 153
Bottle glasses 24
Bottle-making machines 11, 88–91
Bottling jars 102
Brittleness of glass
  factors affecting 22
  research to reduce 189

Bubbles
  as decoration on hollowware **111**, 113
  defects 49
  in manufacturing process 45, 46, 186–7
Building/construction products 108, 168, 172, 173, 194
Bulletproof glass 79, **80**
Bull's-eye window panes 6–7, 10, 54

Carbon amber/yellow coloring 30, 100
Carboys 101
Carl Zeiss Foundation 13–14
Car rear-view mirrors 74, **75**
Cast glass 55
  see also Rolled glass
Centrifugal casting process [for telescope blanks] 184
Centrifuging process 92–3
Ceramizing process 184
Ceran [glass-ceramic] cooking surfaces 107–8
Cerenkov radiation counters/detectors 161, **163**
Chalcogenide glasses 143
Chameleon glass 62
Chemical etching process 72, 75, 117–18, 186
Chemically toughened glass 78–9
Chemical resistance 122
Chromatic dispersion 180
Church windows 7, 55
Classification of glass types 23–6
Coal, as fuel 11, 41
Coating of glass 70, 71, 72, 73, 74, 98
Cold light 176
Cold mirrors 74–5
Colored glasses
  cast/*dalle* glass 55
  compounds added 2, 17, 29–30, 30, 100, 118, 129, 148, 159
  container applications 100, 129
  early manufacture 2, 5
  laminated glass 81
  lighting applications 109, 148–9
  see also Tinted glasses
Color filters 155–7
  narrow-band 157
  steep-gradient 156
Color staining 118, 148–9
Communications technology, optical fibers used 177–82
Composite glass solders 138–9
Composition [of glasses] 24, 25, 133
Concrete, glass-fiber-reinforced 173
Concrete-glass windows 55, **55**, 108
Conditioning 45–6
Container glass 24, 84, 99–102, 194
Continuous processes, tanks used 38–9
Control applications [for optical fibers] 182–3
Cords [defect in glass] 49
Corning special glasses 132–3
  see also Pyrex glass
Corrosion of glass 22–3, 117
CP^2 fibers 180
Crackle glass 113
Crown flint glasses 153
Crown glasses 151, 157
Crucible melts 35, 163
  see also Pot furnaces
Crystal glass 8, 25, 103
Crystallizing solders 138–9
Cutting [finishing] process 115–16

Dalle glass 55, **55**, **56**
Danner [tube-drawing] process 93, **94**
Day tanks 38
Decals [decoration] 120
Decibels 68, 197
Decoration of hollowware 109–20
Defects, melting 48–9
Density [of glasses] *64*, 104, *122*, *129*, *132*
  variation with temperature 153
Dentistry applications 177, 185

Index

Developing countries, glass industries 195
Devitrification 18, 183
Devitrification stones 48
Dewar flasks 98
Diatrate vase 5
Dielectric loss factor *133*
Diopters [units] 157
Discharge lamps 146–8
Discolorations 26–7, 49
Dispersion, spectral 150
Doghouse [batch feeding compartment] 47
Dolomite 28
Dosimeter glasses 162–3
Double glazing 68, **69**, 71, 83
Duran [borosilicate] glass, properties **21**, 104, *122*, 123
Durax [opaque heated] glass 80–1
Dust removal [from flue gases] 191

Economic factors 194–5
Egyptian glassmaking 1–3
Elasticity modulus *64*, *122*, *132*
Electrically heated tanks/furnaces 42, **43**, 192
Electrical resistance, specific *64*, *122*, *133*, 142
Electrode glasses 149, **150**
Electro-float process 62
Electroless metallization 71, 73
Electron-conductive glasses 142–3
Electronics/electrotechnology applications 130–1, 134–49
Enamels 118
Encapsulation of electronics components 134–5
Endoscopy 176
Engraving of glass 75–6, 117
Environmental aspects 190–3
Etched photosensitive glass-ceramic 186, **187**
Etching of glass 72, 75, 117–18
Extrusion, hollowware 92

Facade glazing/walling 70–1, 173
Fiberglass insulation 168–9
Fiberglass-reinforced plastics 172

Fiberglass textiles 169–73
adhesive agents used 171
glass types used 171
processing additives used 171
Fiber light-guides 173–83
assembly of fibers in guides 174–5
cabling 176
image guides 175–6
manufacture of fibers 174, **175**
uses 176–83
Filament inlay 112, **113**
Filament overlay 110, **111**
Finishing processes
cold state 115–18
for hollowware 95–9, 109–20
hot state 110–15
Fiolax glasses 128, *129*, **130**
Fire polishing 117
Fire-resisting glass 82–3
Flame coating 114–15
Flash bulbs and tubes 131, 148
Flashed opal glass 66, 67, 109
Flashing 110–11
Flat glass 51–83
data on soda-lime flat glass *64*
percentage of world production 194
processed products 66–83
production 52–3, 57–9, 60–2
uses 53–7, 60, 62–3
Flint glasses 151, 153
Float glass
applications 62–3
in laminated glass 79
production 60–2
Flue gas dust collection 191–2
Flue gases 43, 191
Fluorescent lamps 144
Foam glass 187
Forest glass [*Waldglas*] 9
Foundry ice [decoration on hollowware] 114
Fourcault process 57, **58**, 62
Fraunhofer, Joseph 10–11
Frosted glass 75, 117, 144
see also Matting of glass
Fuels 9, 11, 41–3
Furnace journey 40

Fused silica/quartz 121–2
  chemical structure **19**
  properties **21**, 121, *132–3*

Gablonzware 98
Gallé, Emile 10
Gaseous emissions 191
Gas-fired furnaces 41, 53
German glass industry 8–9, 10, 11–14, 194
Giftware 102
Glass
  definitions 16–17
  glass-fiber-reinforced 173
Glassblower's pipe 3–4, 85, **86**
Glass-ceramics 106–8, 183–6
Glass charge 33
Glass-crete 108
Glass cullet
  from new melt 30–1
  from recycled/scrap glass 30–3, 101, 192
Glass-fiber-reinforced concrete 173
Glass-fiber-reinforced glass 173
Glass-fiber-reinforced plastics 172
Glass fibers 168–83
  mechanical properties 22, 170
  percentage of world production 194
  see also Fiberglass textiles; Light-guide fibers
Glass modifying components 16–17, 24
Glass silk 169
Glass solders 137–9
Glass spun threads 170
Glass staple fibers 170–1
Glass wool 168
Glassy state
  characteristics 17–23
  materials forming 16–17, 188–9
Glauber's salt 27, 45
Graded index fibers 178, **179**
  manufacture of preform 181
Graded seals 147
Gravity shaft process 167
Greenhouse glass 54
Grinding processes 60, 115

Halogen incandescent lamps 146
Hard glasses 24
Heat consumption [in glass melting] 46–7
Heated glass
  opaque 80–1
  transparent 80
Heat-resistant glasses 25, 63, 123
High-voltage insulators 141
History of glass 1–15
Hoffmann, Josef 10
Hollow structural glass products 108
Hollowware
  finishing processes 95–9, 109–20
  first made 2, 85
  percentage of world production 194
  shaping processes 85–93
  types 84–5
Homogenization 45
Hooke's law 197
Household glassware 84, 92, 102–8
Hydrolytic resistance 102, 129

Ice glass 114
Illax glass 129, *129*, **130**
Image guides 175–6
Image-intensifiers 137, 143
Incandescent lamp bulbs 144
Industrial hollowware processing 96–7
Infrared-absorbing glasses 68, **69**, 135
Infrared lamps 149
Infrared-transparent glass 155
Insulating glass fibers 168–9
Insulating vessels manufacture 97–8
Integrated optics 183
Interferometric control systems 182, 183
Ion colored glasses 155
Ion exchange toughened glass 78–9
Iridescent glass 99, 119

Index

Iron–nickel–cobalt alloys,
   glass-to-metal seals 131, 134
IROX solar reflective glass 71
I-S (Independent Section)
   machine 88, **89**, 105

Jena 2000 glass-ceramic 107
Jenaer Glas products 104
Jena Glassworks 12–13
Jewelry 98–9, 119

Koepping, Karl 10
Kovar-sealing glasses 131, 134
Kunckel, Johann 10
k-values 67, 68

Laboratory glass equipment, 96,
   109, 123–4
Laminated glass 70, 79–81
   in combination with other glass
      81–2, 92
Lamp glasses 143–9
Lampworking 95–6
Lanthanum glasses 153
Laser glass 155, **156**
Leaching, of glass 23
Lead borate solder glass **21**, 138
Lead crystal 10, 24, 104
Lead glasses 10, 24–5, 104, 134–5,
   142
   properties 104, *132–3*
Lead oxides 24, 29
Lens-blank pressing 164, **165**
Libby–Owens process 57–8, **59**
Light-guide cables 176
Light-guide fibers
   manufacture 174, **175**
   uses 176–7
Light guides 174–5
Lighting glass 108–9
Light transmission characteristics
   64, 66–7
Lime 24, 28
Liquid-crystal foil 189
Löffelhardt, Heinz 10
Long glasses 20, 123

Machine blowing processes 88–91
Macor glass-ceramic 185

Marbling effect 114
Marcks, Gerhard 10
Matting of glass 72, 117
Measurement applications 182
Mechanical polishing 60, 117
Mechanical properties 22
Medical hollowware 95, 109
Melting defects 48–9
Melting furnaces/tanks 35–9
   fuels used 9, 11, 41–3, 192
   materials used in construction
      39–41
Melting process 44–50
   alternatives 49–50, 189
   environmental aspects 190–2
Mercury discharge lamps 144,
   146–7
Metallic coatings 70, 71, 72, 73,
   114–15
Micro-blast technique 167–8
Microporous glass 188
Microscope slides and cover slips
   60, 64
Microspheres
   manufacturing process 167–8
   uses and properties 164–8
Millefiori glass 112
Mirror glass
   manufacture 5–6, 73–4
   uses 62, 74–5
Molybdenum-sealing glasses 131
Mosaic glass 111–12
Mouth blowing processes 85–7
Multi-focal spectacle lenses 157,
   158
Multi-lens systems 153–4
Multipane glazing 68, **69**, 81–2
Museums, glass 196

Narrow-band filters 157
Natural glass 1, 16
Neodymium glasses 155
Neon signs 148
Newtonian interference,
   reduction in projectors 72
Nitrogen oxides, removal from
   flue gases 191–2
Non-melting manufacturing
   processes 49–50, 189

Non-reflective glasses 72–3
Nubs 110
Nuclear power applications 140, **141**

Obsidian 1, 16
Oil-fired furnaces 41–2, 53
Olbrich, Josef Maria 10
Opacifying agents 30, 106
Opaline glass 57
Opaque glass 56, 106
Opaque heated glass 80–1
Ophthalmic/optical glasses 12–13, 149–68
  manufacture **38**, 163–4
  properties 149–53
  quality requirements 154
  spectacle glass 157–61
  types 151–4
Optical communications systems 177–82
Optically blown glass 110
Optical position 151, **152**
Origins of glassmaking 1
Ornamental rolled glass 53–4, 77
Osram special glasses *132–3*
Owens, Michael 11, 88

Packaging glass 99–102
  see also Container glass
Painted glass 118, 149
  precious metals used 119
Parison 85
Passivation glasses 139
Paste mold process 88, 90–1, 105
Pasting [of molds] 86
Patterned glass 53–4
Pearls, glass 98, 99
Pharmaceutical glass 127–9, **130**
Pharmaceutical processing equipment 126
Phons [attenuation of sound] 68, 198
Photochromic ophthalmic lenses 159, **160**
Photomultipliers 143
Photosensitive glass 185–6
pH-value measurement 149, **150**
Picture glass 72

Pilkington Brothers Ltd 61, 62
Pittsburgh process 58–9
Plaining 45
Plastics
  glass-fiber-reinforced 172
  microspheres-filled 167
  ophthalmic use 159, 161
Plate glass 6, 60
Polishing processes 60, 116–17
Polychromatic glasses 186
Porous glass 122, 187–8
Potash 24, 27–8
  from plant sources 7, 9, 27–8
Pot furnaces 9, 11, 35–6
  batch feeding 47
  optical glasses cast **38**, 163, **164**
Precious metals
  as thin coating on flat glass 72
  painting on hollowware 119
Preheating 42–3
Preserving jars 24, 102
Press–blow process 89–90, 100
Pressing
  of hollowware 91–2, 105
  of lenses 164, **165**
Primary melting 44–5
Process plant/equipment 124–7
Profile glass 54–5
Projection lamp bulbs 146
Properties of glass 22, *64*, 100
Pulse electro-float process 62
Pyran [fire-resisting] glass 83
Pyrex [borosilicate] glass 63, 104, 123
Pyroceram [glass-ceramic] 107

Quartz glass 121–2
  see also Fused silica/quartz

Radiation attenuation 68, 197–8
Radiation-shielding glasses 72, 161–3
Rare-earth glasses 153, 155
Raw glass 53
Raw materials 26–30
Recuperative heating 43
Recycling of glass 30–3, 101, 192, 193
Reed switches 135, **136**

Refining 45
  in tank furnace 46
Refining agents 30, 45
Reflection
  degree of 154
  increase in **69**, 70
  reduction in 72–3
Reflective flat glasses 73–5
  see also Mirror glass
Reflective foils 165
Refractive index
  lead glasses 103–4
  in optical fibers **179**
  special glasses *122*, **151**, 157, 164–5, 166, 167
Refractive power 157
Regenerative heating 42–3
Retro-reflective foils 166–7
Ribbons, metallic compounds 188–9
Road markings and signs 165, 177
Rocailles [glass beads] 98
Rolled glass 51
  production 52–3
  types 53–7
  uses 53
Roman glassmaking 4–5
Roman glass objects 4
Rotary-mold process 90–1
Roving, spun glass threads 170
Rubber stamps 118, 119

Safety glass 62, 76–82
Safety spectacles 159, 161
Sand
  melting temperature 27
  quality requirements 26–7
Sand blasting 75, 117, 118
Schott Group 14, 63
Schott KPG [tube-drawing] process 94
Schott, Otto 11–13, 66, 78, 151
Schott special glasses *132–3*
Schuller-Metz [tube-drawing] process 93–4
Scientific equipment 94
Scintillation glasses 161–2
Sealed-beam headlights 146

Sealing glasses 130–1, 134–5, 146, 147
Seeds [in glass] 49
Semiconductor glasses 143
Semi-mirrors 74
Sheet glass
  applications 59–60
  see also Window glass
Shimmering colors 119
Shortening point *21*, *64*, *122*, *132*
Shortening temperature, *see* Shortening point
Short glasses 20
Siemens, Friedrich 11
Siemens, Wilhelm 41
Silica *see* Fused silica/quartz
Silicate crystal structure **19**
Silicate glass, chemical structure **19**
Silk-matt etched glass 72
Silk-screen printing 105, 120, 138
Silver-phosphate glasses 163
Single-mode fibers 179–80
Sintered glass parts 139–41
Sinbrax Coffee Machine 10
Snouts 110
Soda ash 24, 27
Soda-lime glasses 23–4
  properties 21, 24, *64*
  uses 51, 77, 84, 103, 109, 141
Sodium discharge lamps 147–8
Softening temperature, *see* Shortening point
Soft glasses 24
Solar radiation reduction 68, 70, 81–2
Soldering glasses 137–9
  composite glass solders 138–9
  crystallizing solders 138
Sol-gel process 49–50, 174
Solid particle emissions 190
Sound-insulating glass products 68, 81, 169
Special glasses 12–13, 25–6, 121–89
  percentage of world production 194
Spectacle glass 157–61
Spectral lamps 148

Spectral lines 148, 150
Spectral transmittance 154
    pharmaceutical glass **130**
Spinning, hollowware 92–3
Stabilizers 28
Stained glass windows 7, 55
Staining of glass 118, 148–9
Steel stamping 119
Steep-gradient color filters 156
Stemware 102, **103**
Stepped index fibers 178, **179**
Stones in glass 48
Strength of glass 22, *64*, *122*
Stresses in glass 22, 76
Striae 48–9
Struck glasses 29
Suck–blow process 88, **90**
Sun protection glass 68, 70, 81–2
Surface treatments
    effect on strength 22
    for flat glass 75
    for hollowware 118–20
Symbols, scientific/technical 196–7
Syrian glassmaking 3, 5

Tableware 25, 87, 102–6
    classification by glass type 103–4
Tank furnaces 11, 37–9
    batch feeding **40**, 47–8
    construction 39
    continuous tanks 38–9, 163
    day tanks 38
    refining process 46
Tektites 1, 16
Telecommunications systems 177–82
Telescope mirror blanks/substrates 122, 184, **186**
Television transmission systems 181–2
Television tube parts 92, 93, 97, 135–6
Tempax [borosilicate] glass 63
Temperature
    ceramizing process **184**
    effect of rapid changes 22, 24, 102

melting processes 36, **37**, 44
soldering 137
viscosity variation 20–1
working 21, 46, 52, 61, 62, *64*, *129*, *132*
Tempered glass 76–8
Thermal conductivity 22, *64*, 67
    see also *k*-values
Thermal–expansion
    borosilicate glasses *122*, *132*
    glass-ceramics 107, 185
    in glass–metal seals 131, 146, 147
    lead glasses 104, *132*
    opal glasses 106
    ophthalmic glasses 157
    pharmaceutical glasses *129*
    and rapid cooling 22
    silica/quartz glass 24, 121, *132*
    silicon 139
    soda-lime glasses 24, *64*
Thermal insulation 68, 70, 81, 169
Thermally toughened glass 76–8
Thermal transfer coefficient 67
    see also *k*-values
Thermometry applications 12, 96, 123
Thermos flasks 98
Thin flat glass 62, 64
Thin-walled containers 100, 101
Tiffany, Louis Comfort 10
Tinted glasses 62, 70, 158–9
    see also Colored glasses
Torch blowing 95–6
Toughened glass
    chemically toughened 78–9
    thermally toughened 76–8
Traffic message signs 177
Transformation temperature 18, **20**
    listed for various glasses *64*, *122*, *129*, *131*, *132*
Trifocal spectacle lenses 157
Tubing-drawing processes 93–4
Tungsten-sealing glasses 131
Types of glass 23–6

Ultrasonic delay lines 141–2

Index

Ultraviolet-absorbing glasses 70, 144, 148
Ultraviolet-initiated crystallization 185–6
Ultraviolet radiators 148
Ultraviolet-transmitting light guides 177
Units, scientific/technical 20, 24, 197
USA
  glassmaking history 14
  production capacity 194, 195

Vacuum coated glass 71, 72, 73, 119
Vanadium-phosphate glasses 142
Vello [tube-drawing] process 93, 94
Venetian glassmaking 7–8
Viscosity 20
  units used 20, 197
  variation with temperature 20–1, 137, 157
Vitreous fused silica 121

Volume, changes with temperature 18, **20**
Vycor 122

Wagenfeld, Wilhelm 10
Waldglas 9
Waste disposal 192–3
Window glass 6–7, 24, 57–60
  applications 59–60
  production processes 57–9
  see also Float glass; Plate glass
Wire compound glass 79
Wire-reinforced glass 52, 54, 82
Wire-reinforced ornamental glass 54
Wood, as fuel 9, 41
Working points/temperatures *21, 46, 52, 61, 62, 64, 129, 132*

X-ray-absorbing glasses 135, 137
X-ray tubes 136–7

Zeiss Foundation 14
Zerodur glass-ceramic **186**